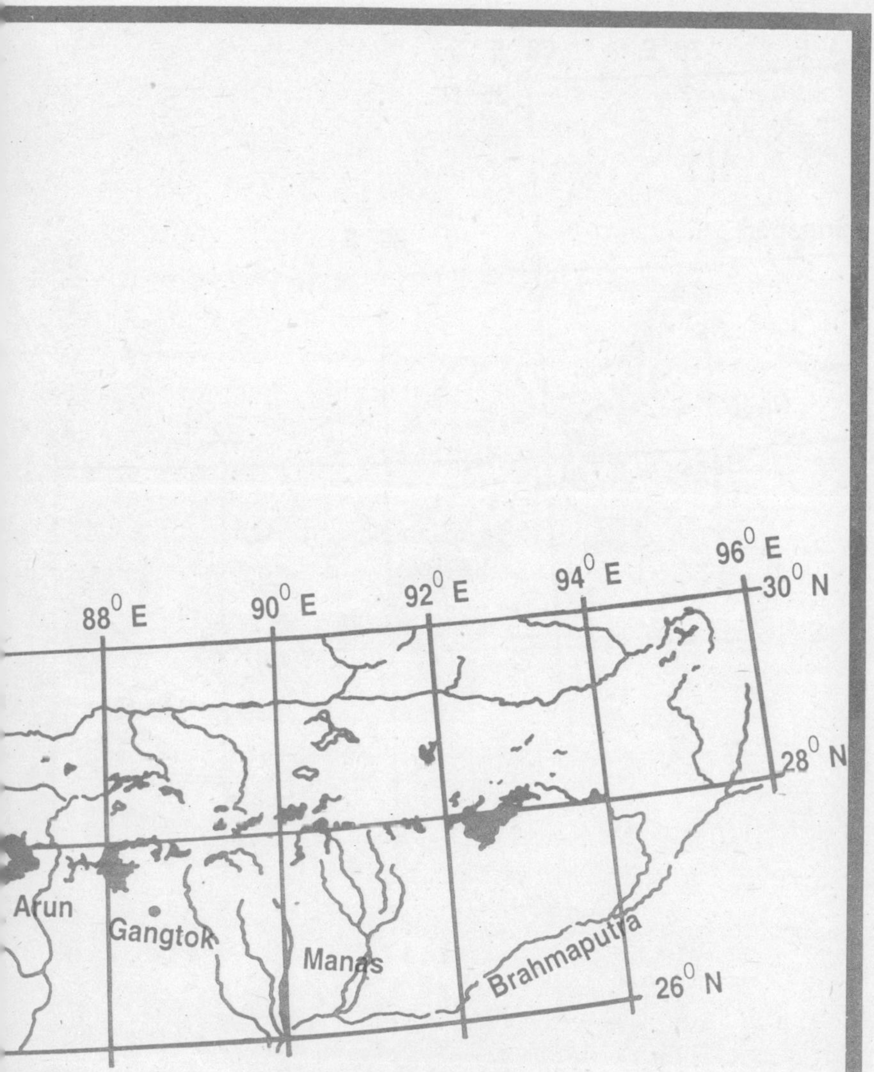

88° E
90° E
92° E
94° E
96° E
30° N
28° N
26° N
Arun
Gangtok
Manas
Brahmaputra

HIMALAYAN GLACIERS: HYDROLOGY AND HYDROCHEMISTRY

According to Puranas:
As the sun dries the morning dew
So are the sins of man dissipated
At the sight of Himalaya

HIMALAYAN GLACIERS: HYDROLOGY AND HYDROCHEMISTRY

SYED IQBAL HASNAIN

ALLIED PUBLISHERS LIMITED

New Delhi • Mumbai • Calcutta • Lucknow • Chennai
Nagpur • Bangalore • Hyderabad • Ahmedabad

ALLIED PUBLISHERS LIMITED

Regd. Off.: 15. J.N. Heredia Marg, Ballard Estate, Mumbai — 400038
Prarthna Flats (2nd Floor), Navrangpura, Ahmedabad — 380009
3-2-844/6&7, Kachiguda Station Road, Hyderabad — 500027
16A Ashok Marg, Patiala House, Lucknow — 226001
5th Main Road, Gandhinagar, Bangalore — 560009
17 Chittaranjan Avenue, Calcutta — 700072
81 Hill Road, Ramnagar, Nagpur — 440010
13/14, Asaf Ali Road, New Delhi — 110002
751 Anna Salai, Chennai — 600002

ISBN 81-7023-922-2

Sponsoring Editor	:	Jyoti Mehrotra
Production	:	B.K. Das
Cover Design	:	Kam Studio

Cover Photograph: Satelite Imagery of Asian Glaciers.

Published by Sunil Sachdev and printed by Ravi Sachdev at Allied Publishers Limited (Printing Division), A-104 Mayapuri, Phase II, New Delhi — 110 064. Typeset by Printo Graphic Systems, Sriniwas Puri, New Delhi-110 065.

Dedicaited to my late father who inspired me to study Earth sciences

ACKNOWLEDGEMENTS

It was in the summer of 1994, during a field trip to Meer de Glace, Chamonix Mt Blanc valley, French Alps and Gornergletscher and Findelengletscher glaciers in the Zermatt valley, Switzerland, that I began to think seriously about the possibility of writing a book on Himalayan glaciers. At that time I was working as a European Union senior 'Marie Curie' fellow in the Alpine Glacier Project, University of Manchester, U.K. While in England I began to collect information on Himalayan glaciers as most of the 19th century research work were published by the British geologists and naturalist in Nature and Royal Asiatic Society. The library of the University of Manchester has been very helpful in providing me reprints of 19th century notes and papers.

I would like to take this opportunity to thank Professor David N. Collins presently at the University of Salford, Manchester, U.K. who has moulded my perceptions at the Himalayan scale.

Special thanks are due to Dr. Suresh R. Chalise, ICIMOD, Kathmandu, Nepal and Professor Gordon J. Young, Secretary General, International Association of Hydrological Sciences, for constant encouragement and support.

Shri Suresh Gopal, Publishing Consultant, Allied Publishers, has arranged review of the manuscript by two anonymous reviewers and comments received from them led to improvement in the presentation of the text.

Finally, acknowledgement is due to my wife Kishwer for her patience and Miss Diana Das for word processing.

FOREWORD

The Himalaya/Karakoram, the largest mountain ranges in the world, stretch across the north of the subcontinent from Assam in the east to northern Pakistan in the west. On the north they are bordered by the Tibetan plateau and on the south by the Indo-Gangetic lowland. Within these ranges lie many thousands of mountain glaciers, the largest of which are up to about 80 km in length and cover up to about 1000 sq km in size.

Major rivers, the Indus, the Ganges and the Brahmaputra originate in these ranges and provide essential water for about 500 million people. About 80% of the waters of the Indus come from melting snows and ice; the contribution of snow and ice melt to the flow of the Ganges and the Brahmaputra is somewhat less, but still very significant. The glaciers themselves provide only a small amount of the total flow of the rivers, but they provide that water in late summer and in years when there is low snowfall. This timing can be critical for water supply for agriculture.

During the 20th century glaciers have, in general, been receding, ice wasting away. Thus they have been contributing extra water to the rivers, over and above that provided by rainfall and snowmelt. However, if they disappear completely, then that extra contribution, often given at times of critical water stress, will no longer be available. This is usually only of small consequence for large rivers, but for many small high-mountain villages highly dependent on glacier melt-water, this may well result in a significant reduction in their water supply. Thus understanding the role glaciers play in the hydrology of these high mountain regions is critical in understanding the nature of the water resource.

Another very significant influence of glaciers on human activity is that of floods. Sudden floods, sometimes of high magnitude, can result from water being ponded back by glaciers and then suddenly released; in other cases landslides released into lakes can suddenly displace large quantities of water which rush catastrophically down-valley causing damage and often bringing death.

Anyone who has seen Himalayan rivers in flood will have witnessed the coloration of the waters indicating high sediment transport. Sediment transport and chemical quality of the waters are of great significance for the variety of uses to which the waters are put in the lowlands bordering the mountain regions.

Thus, the time is right for a book on the hydrology and hydrochemistry of Himalayan rivers. Syed Hasnain is well qualified as author of the book. He has spent many years researching the subject and has gained international acclaim for his work. He has been honoured by the International Association of Hydrological Sciences as one of those younger scientists tasked some years ago to identify the challenges of significance to hydrologists in the decades ahead. He helped in identifying the crucial role which glaciers play in mountain hydrology systems. More recently Dr Hasnain has led a Working Group within the International Commission on Snow and Ice charged with investigating further this role within the Himalayan context.

This timely book, authoritatively written, sythesizes the current state of knowledge on glacier hydrology with its implications for water resource management. It is likely to be a widely referenced text.

September, 1999

Gordon J. Young

Secretary General,
International Association
of Hydrological Sciences

PREFACE

Water-is set to become the major environmental issue of the 21st century. During the coming years problems associated with water shortage will affect the lives of virtually everyone on the planet. Regions of the world facing water shortages are growing in number whilst the demand for water is estimated to have risen six or seven times from 1900 to 1999–more than double the rate of population growth. 70% of our global freshwater supply is locked in glaciers and polar ice. Scientists are predicting that this vital natural reservoir is shrinking at an alarming rate. Water resources of south Asian countries namely India, Nepal, Pakistan, Bhutan and Bangladesh are dependent on the ice melt flows from the Himalaya/Karakoram mountain ranges. The short term consequences of one of the largest and most vanishing glaciers are floods in the valleys whilst the long term effects means drought for 500 million South Asians on the Indo-Gangetic plain beneath the mountains.

A very unique characteristic of the Himalaya/Karakoram, the largest mountain ranges in the world, is the existence of winter-accumulation and summer-accumulation type of glaciers together. For example, the Karakoram, just west of Himalaya, normally belongs to the winter-accumulation type, but in strong monsoon years, a large amount of snow falls, even in summer and similarly strong westerly pressure trough brings snows on the central and eastern Himalaya. This seasonality tends to change from year to year and exert significant impact on glacier variation. Thus variation of glaciers, cause changes in runoff in the water cycle; an increase (decrease) of melting rate on a glacier corresponds directly to an increase (decrease) of river runoff. However, if such condition persists for many years without a change in accumulation rate, the glaciers area where melting occurs decreases (increases), and so the total amount of melt water decreases (increases). In other words, if glacier melt at a given point increases for many years, runoff will eventually reach it's maximum at some time and, thereafter, decreases. Such a tendency in the present day glaciers is attributed to the global warming.

The book provides an overview of important aspects of glacier hydrology and hydrochemistry relevant to an understanding of solute acquisition, sediment transfer, and variations in the runoff. Of particular importance and dealt in considerable detail in the role of monsoon precipitation in the Himalaya/karakoram mountain ranges.

Each chapter in this book tries to introduce the reader to new ideas and concepts not collectively brought together before or synthesized within the field of glacial hydrology and hydrochemistry. Literature is reviewed and placed within the context of present knowledge with the intent to develop a greater understanding of glacial environments in the Himalaya/Karakoram mountain ranges. The aim of this book is not to cover all aspects of glaciology, but only to analyze in detail the quantity and timing of discharge in rivers with snow and glacier melt components of runoff and their quality characteristics during the passage through the subglacial system.

Chapter 1 outlines a brief history of the world and Himalayan glaciers. Glaciers as a source of water resources in the Indian subcontinent is discussed along with its ecological implications. In the last few years it has become apparent that one of the critical parameters in understanding many glacial processes is the hydrology of glacial environments. Chapter 2 explores the many, still poorly understood, aspects of glacial hydrology with an emphasis on dynamics of subglacial drainage system. The chapter gives details on glacial hydrological system with reference to current research in the world. Topics like melt production processes at the glacial surface, mechanics of water passage, movement of water, subglacial drainage system and diurnal and seasonal variations are discussed in great detail with case histories from Karakoram and Himalayan mountain ranges.

Chapter 3 analyzes how meltwater evolves chemically as it routed through a glacier system. Processes of solute acquisition, sources of cations and anions, open and closed system, chemical composition of englacial and subglacial meltwaters, and the controls on the major-ion chemistry of portal meltwaters are also given in detail. The dominance of calcium, bicarbonate and sulphate throughout ablation period in the Himalayan glaciers meltwaters suggests that the chemical weathering of subglacial debris-cover provides significant amounts of sulphate by pyrite oxidation to the high meltwater discharges. The chemical denudation rates in the Himalayan catchments are higher than other glacierized catchments of world.

Chapter 4 outlines the hydrograph separation by using electrical conductivity as a 'surrogate' of total dissolved solids in glacial meltwater. Sulphate as a marker for hydrograph separation has been recently used and analysis of this technique is included and analyzed in the context of Himalayan data.

Chapter 5 examines the characteristics of the sediment system and inter-relationship. While the suspended sediment transport in glacial meltwaters, suspended sediment concentration-discharge relationships, temporal variations and subglacial entertainment of sediment have been analyzed in Chapter 6. Also discussed in this chapter is the conceptually based model of meltwater-sediment interaction at the base of the glacier, relative rates of glacial and non-glacial erosion, and Himalayan and Karakoram perspective of glacio-fluvial sediment transfer processes.

Glacier lakes and their outburst floods (GLOF), a common phenomenon occurring in the glacierized basins in the Pakistan-Karakoram, Indian and Nepal Himalaya is dealt with in Chapter 7. The appendix 1 and 2 gives brief description of methodologies best suited for meltwater discharge measurements, and water quality measurements in the Himalaya-Karakoram mountain ranges.

CONTENTS

CHAPTER 5

CHAPTER 6

CHAPTER 7

1

INTRODUCTION

1.1 GLACIER

A glacier is an accumulation of ice, air, water, and rock debris or sediment. It is large enough quantity of ice to flow with gravity due to its own mass. Glaciers flow very slowly, from tens of meters to thousand of meters per year. The ice can be as large as a continent, such as ice sheet covering Antarctica or it can fill a small valley between two mountains, a valley glacier.

A glacier doesn't start out as glacier. It takes snow on top of more snow, to slowly create a glacier. The snow starts as the beautiful hexagonal flakes as shown in Figure 1.1. The flakes fall to the ground. With time, they lose their edges and slowly become tiny grains, separated by air spaces. As more snow falls, it begins to compress underlying layers. The underlying layers, containing snow grains, become loosely compacted, with randomly oriented ice crystals and connected air spaces. Ice in this form that has survived a season is termed firn. Continued compression removes the air bubbles from firn and form true glacial ice. It has blue tone and takes about 1000 years for snow to turn to glacial ice.

A glacier is a natural body of ice, originating on land and undergoing movement that transport ice from an area of accumulation to an area of disposal. Glaciers can be found at virtually any latitude on the globe where the mountains are high enough and the moisture supply large

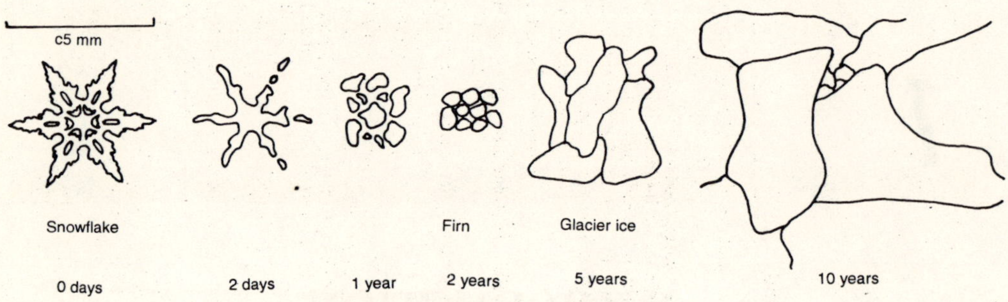

Fig. 1.1. Showing transformation of snowflake to glacier ice

enough to promote permanent ice cover. The presence and absence of glaciers in a basin will depend on many factors like size, orientation and slope of the basin and their interaction with climate. Glacier can be considered as an open system with input, storage, transfer and output of mass. It is a system in dynamic equilibrium where mass balance depends on input and output. Input or accumulation includes all those ways in which mass is added to a glacier: solid precipitation, wind drift snow, and avalanching. Output or ablation include all the ways in which the mass is lost from a glacier: melting evaporation. Figure 1.2 shows a simple model, showing the relationship between accumulation and ablation areas and the way in which the ice moves within the glacier.

1.2 HIMALAYAN GLACIERS

A characteristic feature of Himalayan glaciers is that they receive annual accumulation mainly in the summer by the Indian monsoon. The monsoon precipitation in summer (June-September) occupies about 80% of annual precipitation. Ageta and Higuchi (1984) named these glaciers which have more accumulation in summer than winter, as the 'summer-accumulation type', and differentiated them from the 'winter-accumulation type' glaciers well-known in Europe and north America. On the summer-accumulation type glaciers, accumulation and ablation may occur simultaneously in summer. Under such conditions, summer snowfall sometimes changes to rain under warm air temperature condition, and also new snow cover melts away quickly under such condition. Consequently, summer air temperature in the Himalaya is an important factor which controls ablation through albedo variation at the glacier surface.

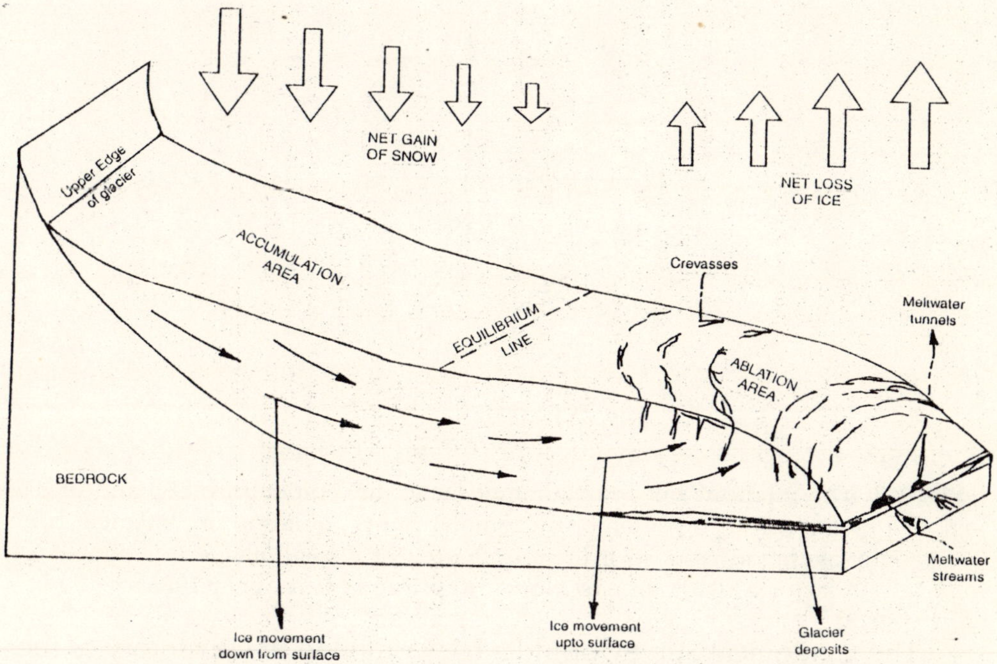

Fig. 1.2. A model of glacier, showing the relationship between accumulation and ablation areas and the way in which the ice moves within the glacier

Thus, summer air temperature is one of the important factors for controlling both accumulation and ablation. Ageta (1983) has estimated relations between summer mean air and accumulation, ablation and balance during summer and these are shown in Figure 1.3 in the area-averaged values for the whole area of glacier AX010 in east Nepal. As seen from this figure, accumulation decreases with temperature rise reflecting increase of rain portion in total precipitation, and ablation increase is intensified by the lowering of surface albedo. As a result of the negative changes of accumulation and ablation both with temperature rise, further decline of balance shows high dependency on summer air temperature.

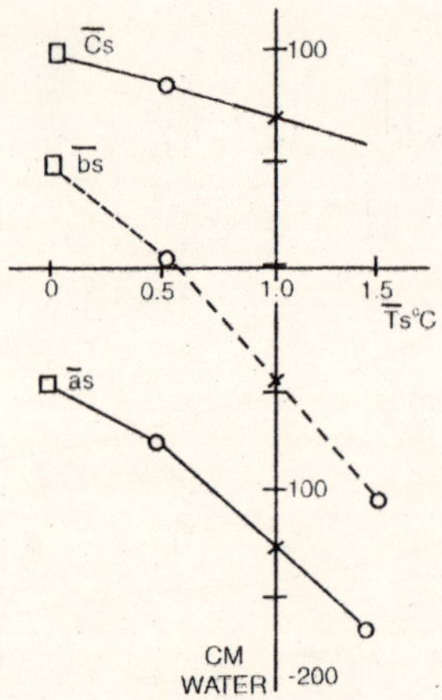

Fig. 1.3. The relations between summer mean air temperature (T_s) and mas balance elements during summer (accumulation: C_s, ablation: a_s, balance: b_s). Spatially-averaged values for the whole area of a small glacier are shown (Glacier AX010), in east Nepal. Reprinted from Ageta(1983)

The Geological Survey of India has compiled the inventory of the glaciers in the India part of Himalaya. Padmashri C.P. Vohra, former Director General of Geological Survey of India under whose leadership the work was done informed me in a 'personal communication' that the glaciers cover an area of 38,039 km^2, broadly divided into three river basins— Indus, Ganga and Brahmaputra on the Indian side of Himalaya. The Indus basin has the largest number of glaciers 3,538, followed by the Ganga basin 1,020 and Brahamputra 662. The glaciers are situated in five states — Jammu and Kashmir, Himachal Pradesh, Uttar Pradesh, Sikkim, and Arunachal Pradesh. Kashmir has the largest concentration with 3,136 glacier covering 32,000 km^2, nearly 13% of the State's territory. The average area of a glacier in the state is 10.24 km^2. 9% of U.P. Himalaya is covered by 917 glaciers having an area of 3,550 km^2. Sikkim has 450 glaciers spread over 912 km^2. The average size is 1.59 km^2. Arunachal Pradesh has 162 glaciers covering 228 km^2. The average size is 1.41 km^2.

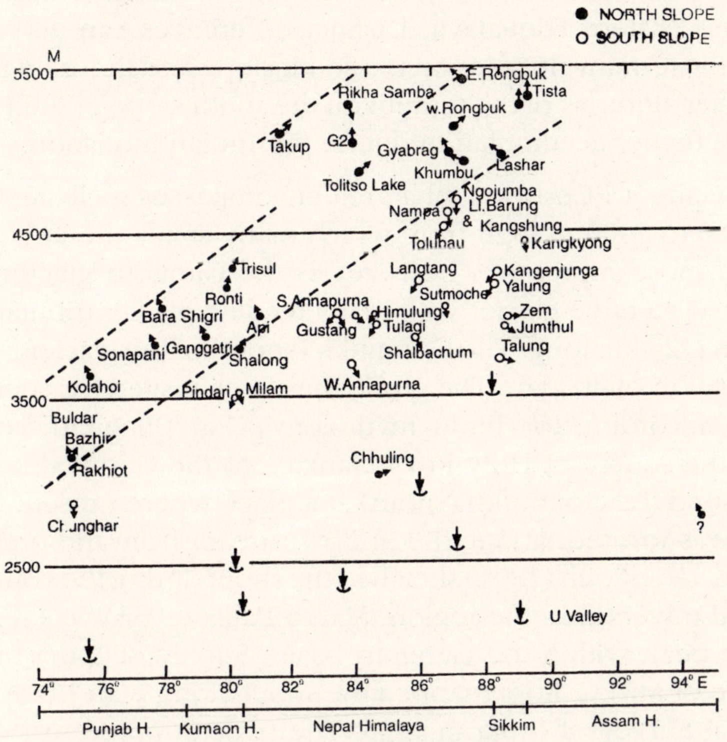

Fig. 1.4. Altitudes of glacier termi in the Himalaya. Reprinted from Fujii (1977)

It has been estimated by various researchers that about 17% of the Himalaya and 37% of Karakoram are covered by glacier ice. The principal glaciers of the Himalaya are Siachen 72 km, Gangotri 26 km, Zemu 26 km, Milam 19, and Kedarnath 14.5 km. Vohra (1993) stated that the glaciers are found in all those areas which attain or exceed the heights necessary for glacier generation. The major clusters of glaciers occur in and around the following ten Himalayan peaks and massifs: Nanga Parbat, the Nanda Devi group, the Dhaulagiri massif, the Everest-Makalu group, the Kanchenjunga, the Kula Kangri area, and Namche Barwa.

Similarly the K-2 region in the Karakoram mountains, Pakistan is highly glaciated and supports several of the largest glaciers. Fujii (1997) plotted the glacier termini from Punjab Himalaya in the west to Assam Himalaya in the east as shown in Figure 1.4. The termini altitudes are low in the western Himalaya as they are located at higher altitudes than those in central and eastern Himlalaya. Distinct differences can also be seen between the termini altitudes on the south slope and those on the north slope: lower altitudes of the termini on the south slope of the Himalaya reflect the higher accumulation due to the Indian monsoon.

The Gangotri-Chowkhamba area encompasses a cluster of many peaks 6000 to 7000 m high from which many small and large glaciers flow. The most impressive, however, is the Gangotri glacier (Vohra, 1996). This is a large glacier with many feeding glacier tributaries. The main stem is 25 km long. The Goumukh (snout) of the glacier is situated at an elevation of 4000 m. The glacier has been visited for centuries by pilgrims. According to Hindu mythology King Bhagirath set out to discover the source of Holy River Ganga into the Garhwal mountains and he found the Goumukh (snout) at a place where present Gangotri settlement is located, about 18 km downstream from the present day Goumukh. Griesbach (1891) sketched the snout during the course of his geological traverses in the region. Macro Pallis (1933) was the first to climb the peak within the Gangotri basin. Survey of India under the leadership of Mr.J.C. Ross mapped the snout along with J.B. Auden, of Geological Survey of India in 1935. The scale adopted was (1:4800). Auden (1935) stated that the glacier must have receded by 740 m during the last century and in the earlier times might have descended to at least Gangotri town and may be even as far down as Jangla. The glacier has been since then visited by various researchers in expeditions by the Geological Survey of India namely, Jangpangi, 1958; Tewari, 1967, Vohra, 1971; Puri, 1972, 1974, 1975, 1976,1977 and 1990. These expeditions focused on the studies of secular movements , glacier deformation, strain rate tensor analysis, ice flow movement, sedimentological studies and the retreat of the glacier termini. Tewari (1967) concluded that the glacier since Auden surveyed it in 1935 had retreated by almost 600 m. Vohra in 1971 took an inter-department scientific expedition on the glacier under the aegis of India Committee for International Hydrological Decade.

The observations conducted by the Geological Survey of India since 1835 have established almost a continuous trend in the glacier recession as shown in Table 1.1. Puri (1990) observed that the most striking recessional period is between 1977 and 1990 when the glacier had vacated 15096 m^2 $year^{-1}$. Whereas, Sahai (1992) has reported that the glacier vacated an area of 0.243 km^2 during the last fifty years (1935-1990). The annual rate of area vacated by the glacier between 1935 and 1971 (36 years) was 8.77% and in the next six years (1971-1997) this value goes upto 10.4%. It is, however, striking that in the last thirteen years (1977-1990) the area vacated by the glacier was exceptionally high 81%. The enormous recession during the last three decades may have resulted by serious ecological imbalances caused by the increased human activity in the Garhwal Himalaya.

Table 1.1 Recession of Gangotri Glacier since 1935

Period	*Annual Ice Cave Retreat (m)*	*Cumulate Area Vacated Since 1935 (%)*	*Annual Area Vacated (Av.) (M^2/ Year)*	*Data Source*
1935 To 1956	10.16 (Along Larger Cave) 4.35 (Along Smaller Cave)	12	2530	Jangpangi
1956 To 1971	27.33	47	12631.5	Vohra (1971)
1971 To 1977	30.84	55.56	5400	Puri (1977)
1977 To 1990	8.082	100	15096	Puri (1990)

The paper by Sahai (1992) has also reported that Geological Survey of India has carried out some glaciological studies pertaining to mass balance on Tipra Bank (1980-1988) and Dunagiri glacier 1984 to 1991 in the Ganga basin.

Mass balance is an important glaciological parameter as the deviations from steady state mass balance conditions cause a dynamic response of the glacier, resulting in a change of flow rate, leading finally to an advance or retreat of the glacier termini. These effects also induce the formation of moraines and other morphological features, which allow delineation of the former extent of glaciers. Unfortunately none of the glacier in the Indian Himalaya has been monitored for long term mass balance observation. In the Nepal Himalaya, however, many, long term observations on mass balance were made by Ageta *et. al.* (1980); Ageta, (1983) on glaciers in east Nepal. They observed on the basis of experimental results on small glacier that the altitude difference between the equilibrium lines for annual balance and balance in summer is less than 50 m due to little accumulation in winter, and the rising equilibrium line is more toward the lower temperature altitudes in the lower precipitation area due to less sensitivity of mass balance to change of air temperature in the colder conditions at the higher altitudes. On the other hand, large glaciers which have large accumulation basins with peaks above 6000 m, precipitation does not change to rain, since it is cold enough to keep snowfall. Therefore, the variations of large glaciers does not depend as strongly on air temperature as that of small glaciers. Inoue (1977) made estimation on mass balance of Khumbu glacier in Mt. Sagarmatha (Everest) region. He showed that the mass supply from the surroundings wide walls by avalanches and snow drifting is more important for equilibrium of mass balance for the whole area of the glacier; such mass supply was estimated to be nearly three times of direct snowfall onto the glacier.

The other glaciers in the Himalaya for which records are available can be divided into longitudinal and transverse types. The glaciers regardless of type, have been in the state of retreat since 1850 (Mayewski and Jeschke, 1979). Lahaul-Spiti glaciers are located in the monsoon arid transition zone (Krenek and Bhawani, 1945). They are extensively covered by debris as indicated by the name of the Bara-Shigri glacier which means'debris covered' in the Spiti dialect (Dutt, 1961). It is a transverse glacier of about 12 km long. This glacier in its retreat process formed a lake between 1860 to 1893 (Egerton, 1864).

The Nanga Parbat massif is characterized by steep topography. Glaciers in this region have experienced comparatively less retreat (6 m y^{-1}) since 1850. Table 1.2 lists the retreat of some important glaciers in the Himalaya since middle of the 19th century. Worldwide the length and mass of glaciers have been shrinking probably at a faster rate than at the end of the past Ice Age. This is perhaps a natural phenomenon, it is possible that an enhanced greenhouse effect could be partly responsible. With the further progress of warming or stabilization of present climate, the glaciers of extrapolar regions of the Earth particularly Asian glaciers will be decaying at rapid, catastrophic rates. Glaciers will survive only in the inner valleys of the Karakoram-Himalaya mountains (Kotlyakov, 1996).

Table 1.2. Principal glacier-fed river system of the Himalaya

River	*Major River System*	*Mountain Area (km^2)*	*Glacier Area (km^2)*	*% Glaciation*
Indus		268842	7890	3.3
Jhelum		33670	170	5.0
Chenab	Indus system	27195	2944	10.0
Ravi		8092	206	2.5
Sutlej		47915	1295	2.7
Beas		12504	638	4.4
Jamuna		11655	125	1.1
Ganga		23051	2312	10.0
Ramganga	Ganga System	6734	3	0.04
Kali		16317	997	6.01
Karnali		53354	1543	2.9
Gandak		37814	1845	4.9
Kosi		61901	1281	2.1
Tista		12432	495	4.0
Raikad		26418	195	0.7
Manas		31080	528	1.7
Subansiri	Brahamputra System	81130	725	4.0
Brahamputra		256928	108	0.4
Dibang		12950	90	0.7
Luhit		20720	425	2.0

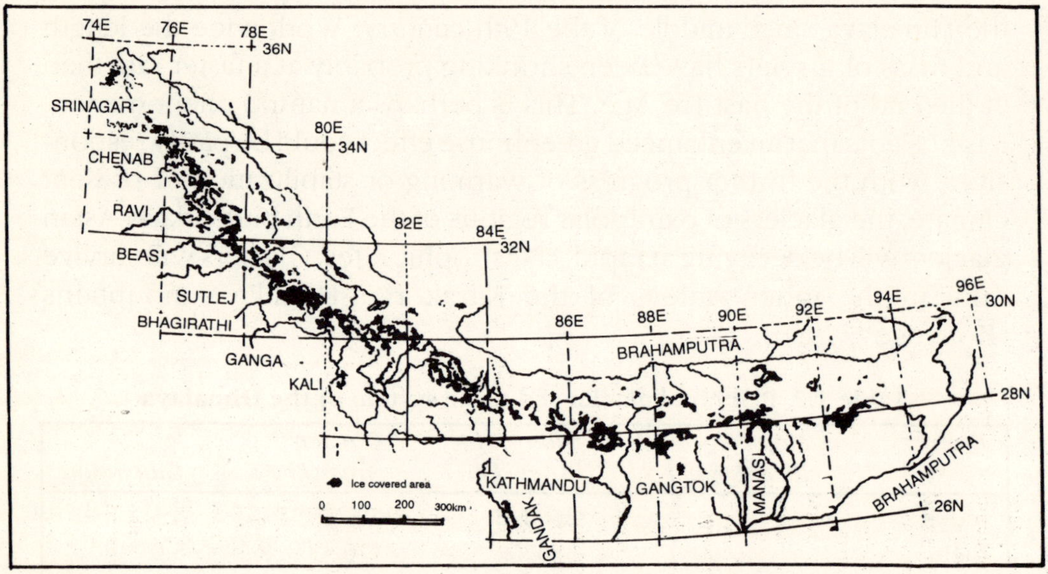

Fig. 1.5. Glacier distribution along the Himalayan arc.
Reprinted from Fujii (1977)

1.3 HIMALAYAN GLACIERS AS A SUSTAINABLE WATER RESOURCE

Glacier hydrology has never been the major focus of the glaciological studies pursued by the Geological Survey of India since late 19th century. It is ironical that snow and glacier covered mountains in the Himalaya are the perennial sources of rivers and streams which flow out of the Himalaya (Figure 1.5). This region is, therefore, immensely rich in water resources and possesses huge potential of hydropower generation. During the last four decades, the population of the Indian and Nepal Himalaya has nearly doubled and the present population of about 120 million people is expected to double by the turn of this century. The growing population's requirement for more food leads to massive deforestation to provide more land for crop production. This exposes the soil, which is then easily eroded by heavy monsoon rainfall, its productivity quickly declines leading to another increase in the demand for more agricultural terraces. At all stages in this downward spiral the water quality and quantity is most severely affected. The natural storage of water in the Himalaya is in the form of snow and glaciers. As a consequence of the possible impacts of global warming and regional and local pollution, the areas under snow and ice are likely to decrease substantially, affecting the low flows in the rivers and hence the supply of water from the Himalaya to the vast adjoining plains of north India. One of the possible impacts of climate change is enhanced monsoonal precipitation in the Himalaya which, in combination with the consequences of deforestation, will further add to increase surface runoff and consequent increase in the incidence of water induced disasters such as floods, mass wasting, landslides, and debris flows (Chalisc, 1994). The climatic fluctuations affect both the amount of snow and ice stored in, and the quantities of meltwater runoff arising from the glaciers. For example much of the stream flow in the Bhagirathi river derives from seasonal snow cover, monsoonal rainfall and the large and small glaciers. Therefore annual meltwater yield is directly related to summer energy input, but the relationship with monsoonal precipitation is very complex. On an annual basis, the runoff maybe reduced when winter snow accumulation has been above average, delaying the rise of the transient snowline and limiting the extent and duration of exposure of underlying ice to melting. Relationships among climatic conditions, glacier mass balance, quantity (volume) of ice stored, changes by advance and retreat of glacier area and runoff are important to the hydrological responses.

The first systematic hydrological studies on proglacial streams were carried out by Kanwar Sain (1946). However, the most noteworthy work was done by Gulati(1973), who delineated the glacierized areas in 20 major rivers having their catchments in the Himalaya (Table 1.3). Besides these studies no serious efforts have been made to investigate the impact of Himalayan snow and glaciers on the river hydrology and climate of north India. The timing and quantity characteristics of the Himalayan snow and glacier runoff is different from those derived from lowland snow or rain . The storage of precipitation in the form of glacier ice over several hundred years can make a large amount of water potentially available, and change the annual distribution of that water. For example, in contrast to basin fed only by snow or rainfall, the runoff from a glacier basin in any one year can be much less or much more than that year's precipitation. The reason for this lies in the importance of the heat balance at the snow/ice surface, especially on the amount of solar radiation and the reflectivity (albedo) of the surface. Hot summers imply higher incoming solar radiation which, together with the heat transferred directly from the air, causes increased melting and runoff, cool summer produces decreased runoff. The amount and extent of snow has an inverse effect on the amount of runoff, as it strongly affects the reflectivity. New snow is highly reflective so that it absorbs less heat and therefore melts more slowly. Old snow firn, and glacier ice have a low reflectivity. Thus, the greater the precipitation in the form of snow, the longer the glacier is covered by a highly reflective material, and the less runoff. A decreased amount of snowfall leads to low-reflectivity surface being exposed longer than usual producing greater melt and increased runoff. Thus runoff from glaciers is naturally regulated and sustainable for human needs.

Table 1.3. Retreat of important Glaciers in the Himalaya

Glacier	*Period*	*Year*	*Retreat in(m)*	*Author*
Milam	1849 to 1957	108	1350	Vohra (1981)
Pindari	1845 to 1966	121	2840	Vohra (1981)
Gangotri	1935 to 1976	41	600	Vohra (1981)
Bara-Shigri	1890 to 1906	16	320	Mayewaski and Jeschke (1979)
Bara-Shigri	1906 to 1945	39	1075	Srikantia and Pandhi (1963)
Kolahai (J&K)	1857 to 1909	52	800	Mayewski and
	1912 to 1961	49	800	Jeschke (1979)
Machoi (J&K)	1906 to 1957	51	457	(Tewari) 1971)
Chungphar (Karakoram)	1930 to 1950	20	600	Mayewski and Jeschke (1979)

The dominance of solid precipitation in the Indus river basin makes the situation somewhat simpler, as the snow and permanent ice provides natural storage while the melting period more or less matches the period of agricultural water need. Whereas the Ganga and Brahamputra rivers have an equal dominance of solid and monsoonal rain precipitation and makes the situation very complex. The melting period coincides with monsoonal period which sometimes causes extensive flooding in the downstream valleys and adjoining plains. However, the monsoonal cloud cover over the entire southern slopes of the Himalaya between July and mid-September cuts off the radiation and substantially reduces the icemelt.

The growing pressure for greater utilization of Himalayan waters derives largely from the strong economic interests of the plains. The possibilities for economic transformation, both within and adjacent to the Himalaya, are enormous if only part of the water volume can be stored, or diverted, and distributed more uniformly in space and time. Any such processes, however, must be integrated with an understanding of the regional environmental and social conditions (Bandyopadhyay and Gyawali, 1994). The International Conference on Ecohydrology of High Mountain Areas, held in March 1996 in Kathmandu, Nepal has highlighted regional issues in ecohydrology, such as data collection and information systems; atmospheric, hydrologic and ecological interactions; and water induced hazards of erosion, sedimentation and landscape processes. The Kathmandu declaration adopted at the end of the conference has recommended that collaborative studies of the hydrological and ecological processes be undertaken along altitudinal gradients taking into account the areal distribution of their parameters in mountainous catchments.

The Department of Science and Technology, Government of India has evolved in 1985 a national coordinated project for the detailed study of glaciers on the southern slopes of the Himalaya. The main objective of this programme is to establish a comprehensive data base on all the important glaciers in the major river basins belonging to various climate zones. One of the thrust areas of the project was to study glacier hydrology and water quality parameters. The information on meltwater yield, its chemical and sediment characteristics is vital to the safety and maintenance of the hydroelectric installations and reservoirs in the outer and inner Himalaya. Recently, it was observed that the Bhakra Nangal reservoir was overdrawing from the ice resources of the Sutlej river by 10 to 13% of the river flow; from this observation it concluded that the rate at which the glaciers are melting in the river headwater may lead to their disappearance in future, destroying the ecology of the area and permanently drying out the perennial rivers (Hasnain, 1989).

The climate and meteorology of the Himalaya control the water resources. If one looks at the entire area from the far eastern Himalaya to the far western Karakoram, one finds that the Indian summer monsoon is the dominant controlling factor for snow accumulation and ablation on the entire southern Himalayan mountains (Figure 1.6). There are regional and vertical differences in the hydrological regimes of the Himalaya with strong seasonality of precipitation. For example in the eastern and central Himalaya the summer Indian monsoon causes snowfall and ablation simultaneously.

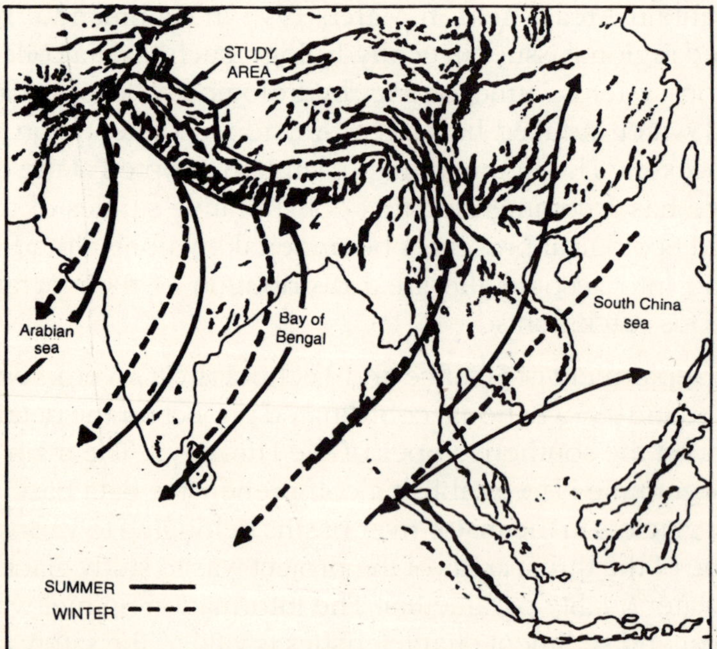

Fig. 1.6. Surface air movements for Indian monsoon. Summer monsoon (solid line); winter monsoon (dashed line)

Glaciers, in the Himalaya, are sustainable sources of fresh water. While every summer large quantities of meltwater flows from them, every winter fresh snow is added. The permanent reservoir of glacier ice is enormous in the Himalaya. Therefore, a detailed data base is required on seasonal and permanent ice cover alongwith establishment of a hydrometric network with measurement of climatic variables over the range of elevations in which the fusion of snow and ice occurs. The principal scientific issues to be addressed are:

(1) What are the seasonal variations in the amounts of runoff derived from the melting of snow and ice ?

(2) How much do the timing and magnitudes of seasonal variations of runoff change from year to year, and in response to what synoptic climatic conditions ?

(3) Which hydrometeorological variables at which measurement stations are best related to runoff variations and this will be used to predict variations in meltwater flows ?

(4) What delay arises within glacierized basins between melt production and discharge at the glacier terminus ?

1.4 HISTORY OF HIMALAYAN GLACIOLOGY

Jean Louis Radolphe Agassiz (1807-1873), a natural scientist from Switzerland was mainly instrumental in glacial theory in its modern form. Earlier to Agassiz many naturalists from Switzerland, France and Austria had talked about the glaciers and the ice movement. In 1837 Agassiz in his presidential address to the Swiss Society of Natural Sciences said "I wish to talk about glaciers, moraines, and boulders. The extension of glaciers is not local but caused by the global climate change which affected the whole of Europe". These observations were published in a book in 1840 titled "Etudes Sur Les Glaciers". The book was widely read in Europe, North America and Britain. Prior to the publication of Agassiz theory the erratic and superficial deposits in Britain were generally attributed to the Deluge (Biblical flood). Prof. A.C. Ramsay, President of London Geological Society stated in his address in 1862, "I am partly convinced of the theory of Mr. Agassiz, that great part of north America, northern part of Europe, and a great part of Britain and Ireland were covered by the sheets of true glacier ice in motion, which moulded and scooped out depressions that subsequently became lakes.

Geologists and naturalists in the British India took interest in the glacial theory as proposed by Agassiz and many papers describing the glacial geology and geomorphology were published. A glacier was discovered in Kumoun Himalaya by E. Madden, in 1847 which he named as Pindari. Subsequently, the geologists of Geological Survey of India have reported various glaciers from the Hindu Kush-Himalayan region. Blanford, published in nature his notes and observations in years 1873, 1877, and in 1891. The glaciers in the Karakoram mountains were brought to notice by Godwin-Austin (1862). The other notable studies on the snow and ice in the Himalaya in the 19th century was done by Oldham (1904), Greisbach (1887), Jack (1844), Hamboldt (1845) Balten (1845), Hilton (1849), and Cunninham (1849).

The Geological Survey of India between 1907 and 1910 had started monitoring glacier fluctuations more regularly. Auden, has published in 1937 a comprehensive report after observing the snout of the Gangotri glacier Tehri-Garhwal since 1934. He has also stated in his report that "with certainty during the Pleistocene Period the Gangotri glacier descended at least as far as just below Gangotri town (located 18 km downstream to the present snout location) it may have descended as far as Jangla". Similarly the glaciers of the neighboring Alaknanda Valley was found to have descended as far as down as Badrinath temple.

First International Geophysical year (1957-58) and later International Hydrological decade (1965-75) have motivated the countries sharing the Hindu Kush-Himalaya (India, China, Nepal, Pakistan, Bhutan and Afganistan) to initiate studies on snow and glaciers. Some of the countries like Pakistan and Nepal have collaborated with countries from North America, Europe and Japan and whereas countries like India and China have started their own research programmes by mobilizing with their own resources.

India:

In India the glaciological studies, were poineered by the Geological of India. The impact of International Hydrological Decade programme had created awareness to initiate systematic scientific studies on glaciers with linkages to climate and hydrology. Many expeditions were launched on glaciers in different climatic zones, but unfortunately none of the glaciers was monitored regularly on long term basis. In 1986, however, a nationally coordinated scientific programme on Himalayan glaciology was initiated by the Department of Science and Technology, Govt. of India. Under this programme Chhota-shigri glacier in Lahaul-Spiti Valley, Himachal Pradesh was studied between 1987 and 1990 and later Dokriani Bamak glacier, in Ganga headwaters, Garhwal Himalaya was studied between 1992 and 1995.

A long term research programme has now been evolved by the Department of Science and Technology, Govt. of India for Himalayan glaciers. The research focus would be on glacier hydrology , mass balance, water quality, sediment transfer, glacial fluctuation, dendroclimatology and remote sensing. Gangotri and Dokraini glaciers, in the Ganga headwaters, Garhwal Himalaya has been selected for long term studies.

China:

Glaciers are distributed on the northern slope of China Himalaya. The area covered by glaciers is about 30,000 km^2 and the total number of glaciers is 11000. Since 1982 important research work has been done in the areas of ice structure, snow cover, snow chemistry and ice cover and glacier mass balance. Mass-balance measurements have been conducted on several glaciers on the Naimana'yi mountain, head of the yarlung Zangboriver, the Xixiabangma mountain, the Gangbo mountain and Qomolangma mountan. In the southern part of Tibetan Plateau, glaciers have suffered great mass loss and shrinkage, however positive mass balance has been observed on glaciers in the north of the Plateau. The total annual glacier meltwater in China is 564×10^8 m^3. The glaciers play an important part in annual runoff change in western China. The snow resources are very abundant and are mainly in the north west, north east and Tibetan plateau. The total area of snow cover is nearly 1×10^6 km^2. The water equivalent is about 100 billion m^3. The snow and glacier studies are mainly done at the Institute of Glaciology and Geocryology, Lanzhou, Ganshu province.

Nepal:

Survey on glaciers was initiated in 1899 by D.W. Freshfield in Kangchenjunga massif. Burrad and Hayden also surveyed in this region in 1933. In Mt. Everest (Sagarmatha) region, N.E. Odell made preliminary survey in 1925. In 1956, Swiss expedition under F. Muller surveyed the movement and regimen of Khumbu glacier.

Professor Higuchi group from Nagoya University, Japan did initiate research activities on Nepal Himalaya glaciers by mounting scientific expeditions in 1973, 74, 75,76 and 78. The Department of Hydrology and Meteorology of His Majesty's Government of Nepal launched a snow and glacier Hydrology Project in 1987. In the course of this project, the snow and glacier hydrology unit had established four permanent stations in the high Himalaya namely Langtang, Khumbu-Everest, Annapurna, Makalu and Kanjiroha regions during the period 1987-1990. Recently the Glacier Research group from the university of New-Hampshire, USA has also initiated a project to drill ice-core. The core studies would provide valuable data on regional and global scale climate change.

Pakistan:

The Karakoram glaciers have attracted attention right from the middle of 19th century before the partition of the Indian sub-continent. Many of these glaciers are known for their advances (Desio, 1954; Kick, 1958; Hewitt, 1969). During the International Geophysical year 1957-58 a joint expedition was launched under the British Royal Society. In the recent times, however, a joint cooperative project on snow and glacier hydrology was initiated in the upper Indus basin catchment to study the impact of climate change on water resources development. Professor Gordon J. Young, cold Regions Research Centre, Wilfrid Laurier University,

Canada and Professor David N. Collins, Alpine Glacier Project, Department of Geography, University of Manchester, UK have conducted detailed studies on Batura glacier in upper Indus basin. The focus was on hydrology, glaciology and water quality measurements. The studies were conducted in collaboration with WAPDA (Water and Power Development Agency, Pakistan).

1.5 WORLD'S GLACIERS

The coordinated collection of information about glaciers was begun in 1894, with the purpose of clarifying the relationship between glaciers and climate, and improving understanding of Ice Ages. Until about 1950, most of the information collected related to glacier lengths, particularly in the Alps, Scandinavia, Iceland, Canada, Patagonia and Andes.

Direct mass balance measurements began after second world war, first in the Scandinavia and then elsewhere, many in conjunction with hydroelectric schemes. It was their work that established the link between climate change and glacier fluctuations. The need for a worldwide inventory of perennial snow and ice masses was first considered during the International Hydrological Decade operated by UNESCO during 1965-74. The International Commission on Snow and Ice (ICSI) was asked to prepare guidelines on completing glacier inventory data, and these were produced in 1970. A comprehensive world Glacier inventory: status 1988 was published in 1989. Its aim was firstly to provide a snapshot of ice conditions on Earth during the second half of 20th century. Glaciers inventories were made worldwide, and then compiled to provide a statistical basis for the study of glaciers. Figure 1.7 shows all the important glacierized areas of the world and now covered by some kind of inventory, most of them detailed but preliminary in regions like Hindu Kush-Himalaya and Andes.

The area covered by glaciers worldwide is estimated at 15.8 million km^2, an equivalent of 10 percent of the total land surface. Antarctica has a share of 13.6 million km^2, Greenland 1.7 million km^2. The remaining 3.4% are distributed over the rest of the globe as mountain glaciers and small ice caps; the area covered by some of them are as follows:

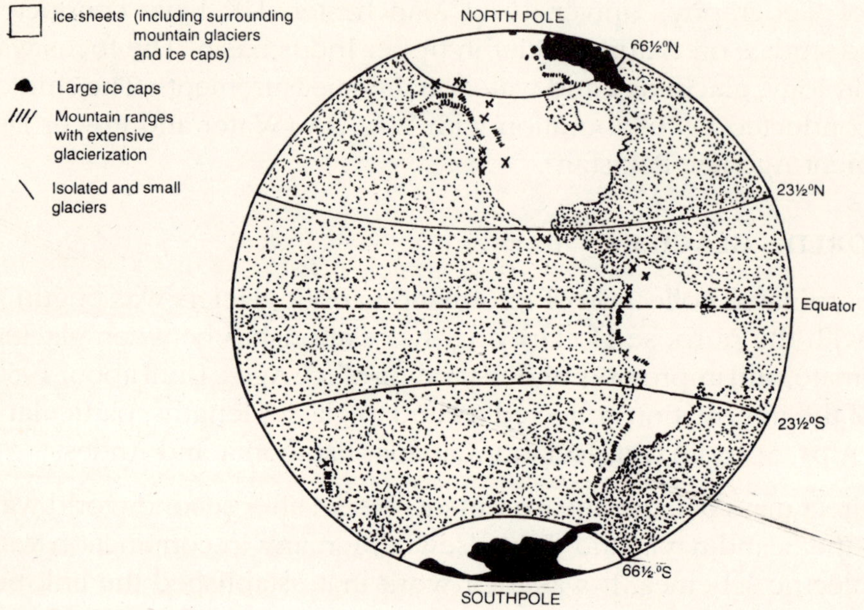

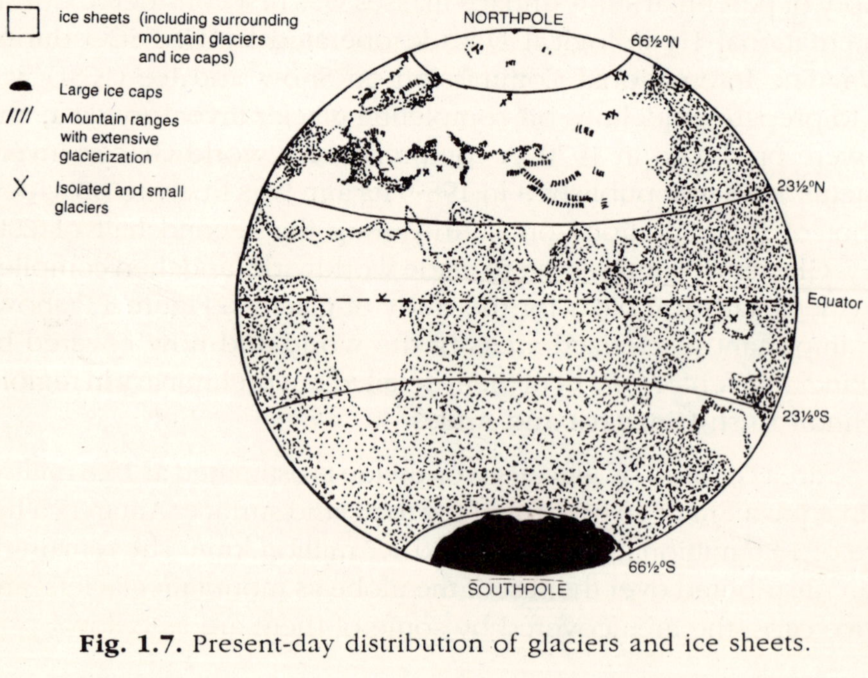

Fig. 1.7. Present-day distribution of glaciers and ice sheets.

Asia	Area (km^2)
1. Commonwealth of Independent States	77,223
2. Turkey/Iran/Afghanistan	4000
3. India/Pakistan	40,000
4. Nepal/Bhutan	75,005
5. China	30,000

Today there are about 30 million km^3 of ice on our planet. In winter snow covers about 66 percent of the land surface of the world (Katloyakov, 1997).

Global land-ice monitoring from space (GLIMS) is an international project started in 1998 is to ascertain the extent and condition of world's glaciers with the accuracy and precision needed for the most accurate possible predictions of future change. This objective supports global and regional climate-change. GLIMS entails: (1) comprehensive multispectral and stereo image acquisition of land-ice on an annual basis, including repeated acquisition during the summer /early autumn seasons of each hemisphere to minimize cloud and snow cover in the season's best images; (2) use of satellite imaging data to measure inter-annual changes in glacier length, area, boundaries, and snowline elevation; (3) where possible, measurement of glaciers surface velocity fields; (4) development of uniform digital Geographical Information System; (5) validation of the GLIMS glacier database through cross checking results with field data. This inventory and global image archive will be useful for assessment of glacier impacts of global climate change at the start of the 21st century and for prediction of future glacier and sea level trends. Most importantly, these records will comprise a uniform baseline against which subsequent glacier change can be assessed as Earth increasingly feels the bite of global warming.

1.6. GLACIERS AND CLIMATE

Glaciers of the mid latitudes and tropics, on the other hand, have comparatively little effect on the atmosphere, their interaction with climatic elements is restricted to their immediate surroundings. The size of an extrapolar glacier is too small for the establishment of a true feedback to the atmosphere: the scale typical for synoptic processes is 10^3 km, compared to the size of larger alpine glaciers of only 10 km. Thus, the vertical fluxes of energy and mass from a glacier can contribute only negligibly to the total exchange between the atmosphere and the Earth's surface, the storage capacity of glacier for these quantities is much larger than that of the atmosphere. A glacier can store a 100 m column of water compared to 1-2 cm of precipitable water present in the atmosphere in mid latitudes. An amount of 10^7 J will heat a 1-m^2 column for out entire atmosphere or only about 5 m of ice by 1°C or will melt 0.03 m of ice.

It is understandable that this high storage capacity for mass and energy will make the glacier respond slowly to the fast changing state of the atmosphere. A glacier acts as an integrator of meteorological processes in a variety of time scales: the surface snow cover and the adjustment of the equilibrium line has a yearly cycle. The time for flow adjustment and size changes has a scale 1-10 years (or 100 for large glaciers), and finally, 100-1000 years is the order of magnitude of the average residence time of ice in the glacier. These three time scales can be applied to three climatological problems:

(a) the immediate reaction of the glacier surface to the climatic variables,

(b) the dynamic adjustment of the glacier to climatic fluctuations trends for several years,

(c) the archives of secular changes in atmospheric composition contained in the ice and its air inclusion.

A glacier is a very suitable tool for the measurement of atmospheric conditions; from a study of present day coupling between glaciers and their environment, we should be able to retrieve the information about past climates that is stored in glacier and in the morphology of its surroundings.

This optimism, however, is damped by the fact that the signal from the immediate vicinity may be of a similar magnitude to that from the atmosphere. In extreme cases, the system will undergo irreversible changes since the glacier response is faster than that of morphological or botanical adjustments. For instance, a glacier may advance through grassland, but when it retreats its terminus will be covered and surrounded by debris.

The study of the climate-glacier relation is relatively young: 50 years ago the Scandinavian school applied micrometeorology to the glacier surface while the first theories of ice flow were being tested. The development of glacial meteorology has been reviewed by Hoinkes (1964, 1970) and Radok, (1978), and a comprehensive list of energy and mass budget studies up to 1966 has been compiled by Paterson (1996). Meier (1965) summarized the state of the art by chain of processes which combined the climatic and the dynamic responses that lead to the morphological evidence of past variations. From this chain (modified from Meier, 1965):

(1) Accumulation ↘	↙ (2) Ablation	Meteorology
(3) Mass budget		Glaciology
(4) Dynamic response		
(5) Position of moraines		Morphology

can easily identify the fundamental difficulties inherent in the reconstruction of past climates from glacial morphology. Although the dynamic behaviour of a glacier is basically understood (Nye, 1961; Paterson, 1996; Budd & Jenssen, 1975), it is difficult to assess the response for an unmeasured glacier (Budd & Allison, 1975). This problem can be overcome in practice by analogue and by empirical parametrization of glacier flow.

The reaction of a glacier to small climatic variations can be described by a combination of energy and mass budget. The behaviour of the equilibrium line is predictable from accumulation, radiation budget and air temperature and their vertical gradients. These variations are nearly equal to the standard deviation of long term yearly fluctuations of the respective variables. As they are not mutually independent it is most likely that a combination of the three will occur. The vertical profile of the mass budget shows little variation from its equilibrium shape when positive or negative budget situations are inspected. Small as it is, this variation is systematic and shows the importance of the glacier tongue for negative budgets while the budget at the tongue does not vary much from balance to positive years.

Climatic trends are clearly reflected in mass and temperature changes in glaciers and permafrost. It is for this reason that perennial land-ice bodies are key parameters for climate system monitoring (Haeberli, 1990).

Today's strategy of worldwide glacier monitoring is an attempt to combine various types of information. Observations of glacier length changes remain the essential key to the past and the most practicable possibility of reaching global coverage. Haeberli (1998) explained that the quantitative interpretation of glacier length changes with respect to mass balance and climate change relies on what is learnt from the limited number of available detailed and long-term mass balance measurements in different climatic regions of the world. Such available long series of extensive mass balance measurements can be done using index stakes (Linear balance model) combined with repeated mapping. This approach gives the mean mass balance and thus the direct glacier signal of climatic change on a regional scale.

With the accelerated warming, large glaciers would continue downwasting rather than retreating. Repeated mapping or profiling with a combination of laser altimetry and global positioning system (GPS) would give important information in such cases, especially with regard to meltwater production. Systematic application of advanced remote-sensing and modeling techniques indeed appears to be the main challenge for worldwide glacier monitoring into the 21st century.

1.7. THE FUTURE OF HIMALAYAN GLACIERS UNDER THE EXPECTED CLIMATE WARMING

Retreating terminii of several Himalayan glaciers have been observed since the middle of 19th century in the Kashimir, Kumoun, Sikkim, Garhwal and Nepal Himalaya. The retreating speed of some of the glaciers has accelerated since the 1980's (Yamada, 1992). Data on glacier length change (retreat) are regularly reported from Pakistan, India and Nepal (IAHS(ICSI)/UNEP/UNESCO, 1993).

The global circulation models and Intergovernmental Panel on climate have established that the rapid increase in the concentration of green house gases has caused rapid rise in global and local temperatures. A twofold increase in the content of atmospheric CO_2 may result in global temperature increase by 2-4°C; such conditions are predicted by the year 2030 (Kotlyakov, 1996).

Variations of the summer-accumulation type glaciers in the Himalaya are sensitive to summer air temperature , solar radiation and precipitation. Shrinkage of glaciers caused by global warming may lead to a decrease in water resources and to an increase in hazards from outbursts of glacier lakes. Thus, snowmelt runoff will decrease dramatically, with the peak of snow melting shifted to the early spring period. The extrapolar glaciers of the Earth will be decaying at rapid, catastrophic rates—and its total area will shrink to half by the year 2050 according to the analyses done by Kotlyakov (1996). He further opined that the glaciers will survive only in the mountains of inner Alska, on some Arctic archipelagos, within Patagonian ice sheets, in the Karakoram mountains, in the northern Himalaya, and in some regions of Tibet. Consequently, the glacier recession as a result of the 'green house' effect will develop highly negative mass balance (–3 or even –5 m per year). Therefore, runoff from glacier fed rivers like Ganga, Indus and Brahmaputra will drop dramatically causing negative impact on agricultural production in the countries of south Asia.

Cryosphere models suggest that between 1/3 and 1/2 of existing mountain glacier's mass could disappear over the next 100 years. The reduced extent of glaciers and depth of snow cover also would affect the seasonal distribution of river flow and water supply for hydroelectric and power generation and agriculture. Anticipated hydrological changes and reduction in the areal extent and depth of permafrost could lead to large scale damage to infrastructure, an additional flux of CO_2 into atmosphere, and changes in the processes that contribute to the flux of (CH_4) into the atmosphere. However, little change in the extent of Greenland and Antarctic ice sheets is expected over the next 50-100 years.

In Venezuela, three glaciers have disappeared altogether since the early 1970's. Three have disappeared from Mount Kenya in Africa. Every glacier on which have any data shows a very rapid retreat taking place.

The vanishing mountain glaciers, permafrost and snow cover are an early sign of manmade global warming caused by the carbon dioxide we have added to the atmosphere. This might be enough to evaporate a lot more water off the tropical ocean. The water vapour might then amplify the warming enough to melt the glacier ice.

1.8. DEBRIS-COVERED GLACIERS

There are debris-covered glaciers, sometimes called moraine-covered or supraglacial covered glaciers in many regions of the world (Kotlyakov, 1996). Regions with high occurrence are the Himalaya Karakoram, southeast Tibet, and New Zealand. These are glaciers in which debris are accumulated in the ablation area. These regions are located at the periphery of the tectonic plate of the Earth's surface where it corresponds to subduction zone or site of frequent deep earthquakes, which enhances debris supply to the glaciers surface mainly in the accumulation area.

Large glaciers in the Indian, Nepal, Bhutan Himalaya and Karakoram mountains are covered with thick supraglacial debris in their ablation area. The debris affects glacier ablation process and the runoff amount. Morphology and structure of the debris and its dynamically effect also characterize the glacier response to climatic variations in comparison with debris free glaciers (Ageta and Pokhrel, 1999). From statistical analyses of glaciers in Khumbu region by Fujii and Higuchi (1997) observed that the average area and length of the debris covered-glaciers are 15 times larger and 5 times longer than those of debris-free glaciers respectively; 35 % of the total glacial area and 42 % of the total area of debris covered glaciers are covered with debris. Such large extent of the debris and debris-covered glaciers indicates importance of those in the hydrological process and water budget in large glacierized basins. The debris on the Khumbu glacier was determined area by from rock types to be supplied from the rock walls surrounding the accumulation area of the glacier as shown in Figure 1.8 (Fushimi *et al.*, 1980). Average daily ablation (water equivalent) in August was 2.5 cm with site to site variation in the upper half of the ablation area (6-9 km from the terminus) where glacier flow was active, while in the lower stagnant part (0-4 km from the terminus) was less than 0.4 cm due to insulation effect of thick

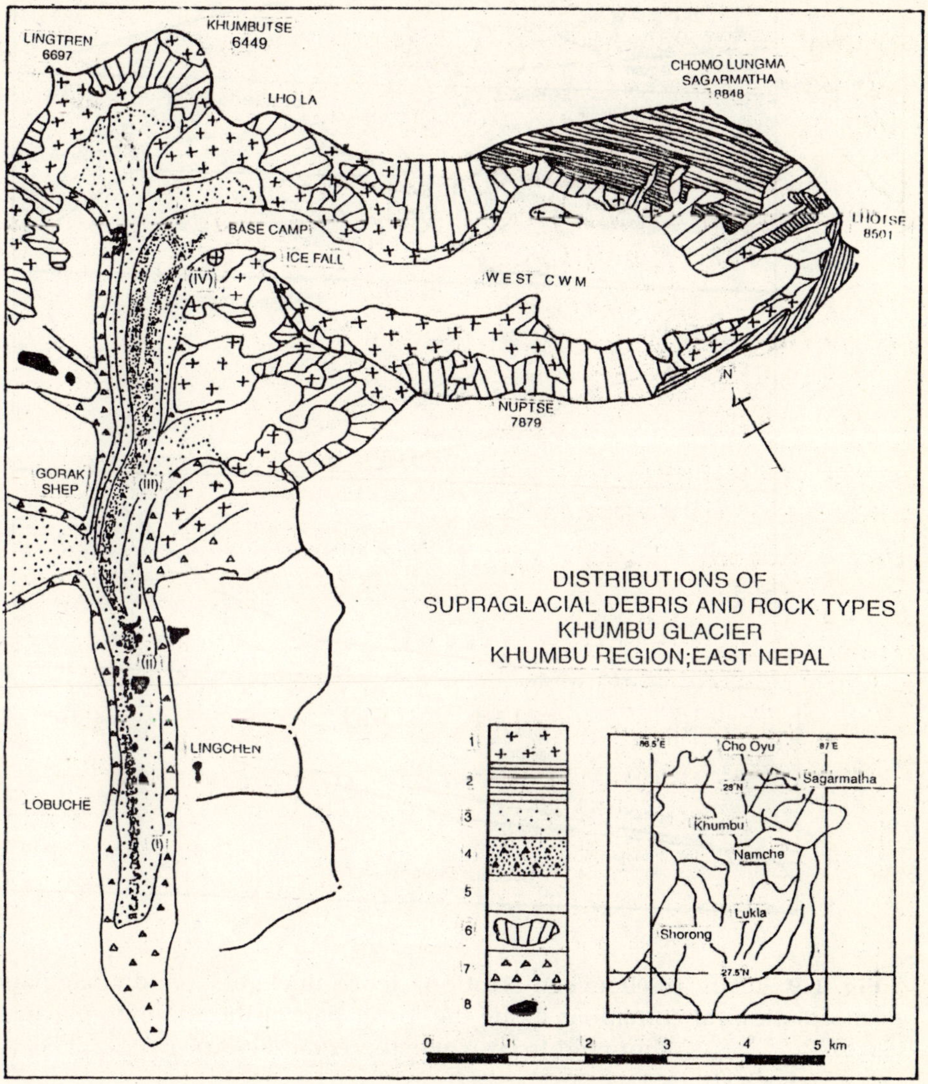

Fig. 1.8. Distributions of supraglacial debris and rock types Khumbu glacier, Khumbu region, East Nepal. Reprinted from (Fushimi *et al.*, 1980)

debris cover more than 1 m (Inoue and Yoshida, 1980). Figure 1.9 shows the relations between ablation, debris thickness, ice drainage and morphological conditions along the lower part of Khumbu glacier (Watanabe *et al.*, 1986).

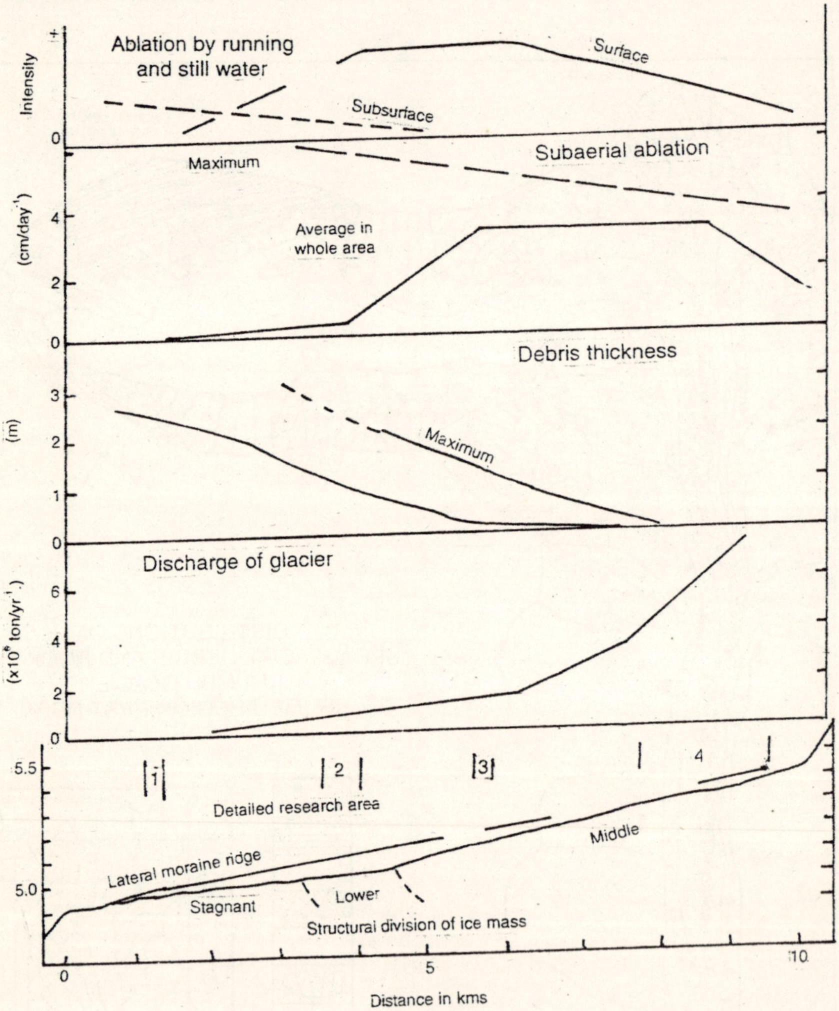

Fig. 1.9. Relatio1ns between ablation, debris thickness, ice drainage and morphological conditions along the lower part of the Khumbu glacier. Reprinted from Watanabe *et al.*, (1986)

A thin debris layer on a glacier intensifies glacier ablation, because the albedo of debris is much lower than that of snow and ice; effect of supraglacial debris on glacier ablation is changed by the thickness of the debris (Fujii, 1977). The effect of the debris on ablation, in combination with glacier dynamics, produces complex surface features with irregular uneven topography in the ablation area such as debris-covered cones, ice cliffs, hollows, ponds and lakes. And thick debris cover (Figure 1.10) keeps fossil ice beneath it in the stagnant part of glaciers which attaches to the active glacier body in the upper stream. Role of the supraglacial debris in the Himalaya and Karakoram mountains are very complex as they influences the sediment and solute acquisition processes during the months of monsoonal precipitation (Hasnain and Thayyen, 1999).

Fig. 1.10. Debris cover lower ablation region of the Gangotri glacier, Garhwal Himalaya

2

GLACIER HYDROLOGY

2.1 INTRODUCTION

Glaciers, including the Antarctic and Greenland ice sheets, ice caps and valley glaciers in polar regions and in high mountains, retian significant amount of water on land, and they are the largest fresh water reservoir. They contain more than 75 % of the total fresh water on the surface of the earth.

Water occurs as a fluid, as an interstitial liquid, as a solid and as interstitial ice within all glacial environments. As meltwater, it is active in debris erosion, entrainment and deposition. Meltwater causes channel bed erosion and scour, and the development of complex drainage networks within all glacial environments (Menzies, 1995). Glacial hydrology, therefore, is an integral element in understanding all glacial environments and processes. According to the world Atlas of Snow and Ice (Kotlyakov, 1997) there are about 30 million km^3 of Ice on our planet. Thus in some aspects we are still in 'Ice Age'. In winter, snow covers about 66% of the land surface. The total glacier-derived runoff, equal to 3,450 km^3 of water, accounts for about 8% of the total surface water runoff to the ocean (44,700 km^3) and for about 3% of the land precipitation (119,000 km^3). Of this amount, Antarctica and Arctic Islands contribute 3,010 km^3, and mountain glaciers 440 km^3, or 1.0 % of the land surface runoff. Nearly another 1% (370 km^3) is contributed by the seasonal runoff of glacier snow in summer periods in the absence of melt-water runoff from ice-free areas; the latter factor is very important in enhancing the value of glacier-derived runoff as a water resource (Kotlyakov, 1996).

The hydrology of glacierized regions is thermally controlled. Runoff results from interaction of precipitation with environmental thermodynamic characteristics. Variations in energy availability lead to fluctuations in melting of snow and ice and the production of meltwater. Because of the thermal threshold snow and ice masses are prevented from entering the liquid phase until the critical melting temperature has been attained. Seasonal variations in the form of precipitation from winter snowfall to summer rain, and of energy supply usually peaking to a summer maximum, produce strong seasonal periodicity of hydrological event which influences quantity, quality and timing of drainage (Young, 1985).

Solid precipitation enters temporary storage and liquid in the drainage system. The surface meltwater also enters the drainage system. The structure of the drainage system and various storage within the ice-mass and at the sole of the glacier controls the basin response. A glacier is a fixed location feature in the hydrological cycle of the basin, but the snowcover which covers both the ice and ice-free areas in winter months is transitory, and variable in aerial extent and spatial distribution.

The high mountain hydrological system may be subdivided into glacier-covered and ice-free zones. In some catchments of high relief permanent ice may be absent, as every year all the accumulated snow melts during summer . In other mountain basins some snow will persist from one year to the next, at higher elevations nourishing glacier-cover over a portion of the basin area. Whereas in the remainder almost all the snow will melt, leaving a few residual snow patches, which may persist through several years. The principal differences in behaviour between nival (ice-free) and partially glacierized basins are that glacier ice melt provides a source of runoff throughout summer, from the time at which snow first clears from the glacier tongue, and that the lower albedo of ice, by comparison with snow leads to higher melt rates. Figure 2.1. shows various hydrological elements of a typical Himalayan valley glacier. The precipitation vary, between mountain ranges within mountain ranges, seasonally and from one year to another, types of precipitation also vary; the freezing elevation is not constant and thus the proportions of rain and snow from place to place and through time vary considerably.

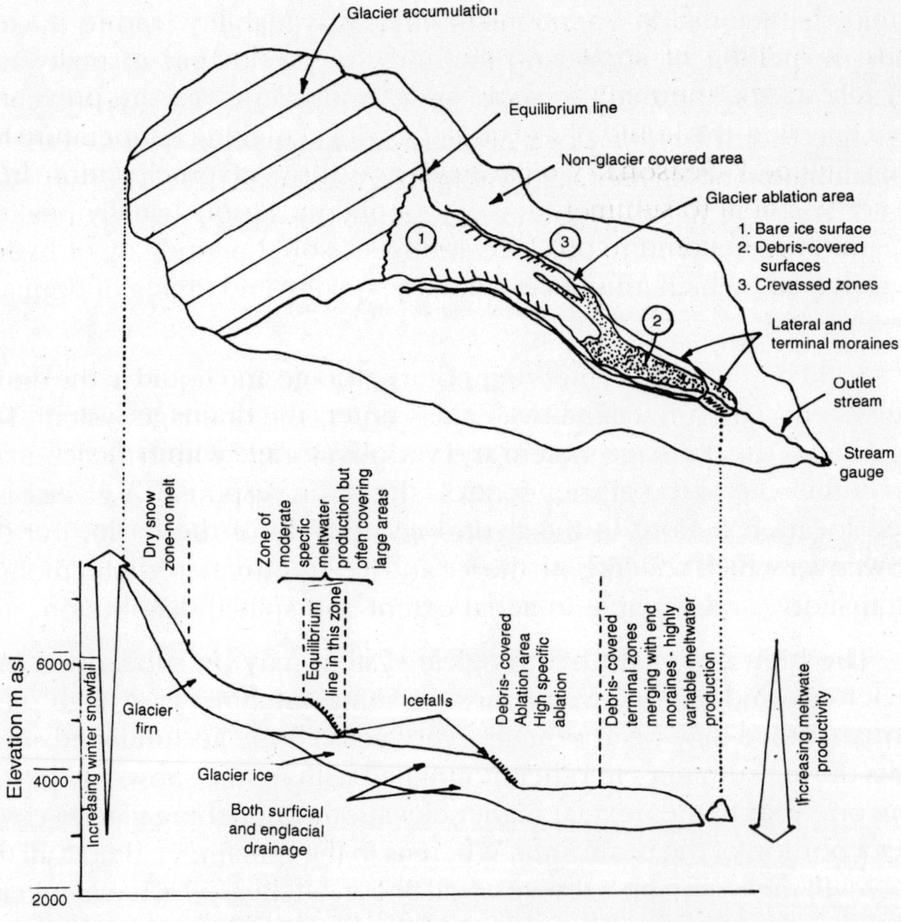

Fig. 2.1. Hydrological elements of a typical Himalayan glacier. Reprinted from Young and Hewitt (1992)

In high mountain areas, over an annual discharge cycle, specific runoff (discharge derived from a unit area of catchment) is generally higher than in surrounding plains because of greater precipitation inputs and reduced evaporation. With increasing elevation of basins, specific runoff is usually enlarged. This statement must be modified for most glacierized regions from which specific runoff is smaller as they behave like natural reservoirs. However, the distinctive characteristics of high mountain runoff relatė to the timing of flows. Because some of the precipitation accumulates in snowpack during winter, at temperature below the melting point of O°C, much runoff is delayed from the time of precipitation until later in the year. Not only is the solid form of precipitation retained, but rain falling over a snow cover is absorbed and retarded. Flows of streams and rivers in mountain are therefore concentrated in spring and summer months, when water is released by melting from snow and glacier ice. By contrast, winter discharges are low or negligible, according to the duration of temperature below freezing, controlled by the frequency of fluctuation of the freezing level. Mountain runoff, therefore, largely reflects change in heat energy available for melting. Whether the melt is derived from seasonal snowcover alone or from a basin also containing perennial snow and glacier ice. Runoff is a hydrological response to variation in thermal conditions.

Nival basins maintain no perennial snow and ice, but receive seasonal snow cover. Runoff from snowmelt in mountainous nival basins occurs later in the year than that at lower altitudes. Lower air temperatures lead to colder winter snowpack, which require progresses and snowpacks disappear, this lag effect is reduced, runoff from glacier ice surfaces is fast, and diurnal fluctuations are more pronounced following the diurnal fluctuation in available energy for melting.

Typically, minimum flow of a glacierized river, close to the glacier, is recorded in the early hours of the morning, between seven and ten , and the maximum reaches in late afternoon, between three and six, so that each diurnal hydrograph is asymmetrical, with the rising limb steeper that recession, and the peak lagged by several hours after the time of maximum solar radiation. Although nival rivers do have as marked rhythmic diurnal hydrograph component, daily flow ranges are small, since input characteristics are modified as meltwaters percolate through snowpack. Lag is considerable, and there is a strong contrast between nival and glacial diurnal hydrographs.

The seasonal variation of runoff from a mountain basin bears little relation to that of precipitation. Essentially a long lag exists between the accumulation of precipitation in winter and its release from storage during summer in both nival and glacial catchment areas, with the effect becoming more extreme with increasing glacierization. On a daily basis, relationship between precipitation and runoff are complex and depend on the form of the input and on conditions at the surface. Snow in winter builds up passively, whereas in summer snow decreases runoff because increased albedo reduces melting. Rainfall is stored by snowcover from which it is released gradually or retained as ice according to thickness, density and temperature of the pack, whereas in summer, when ground is free from snow, and in glacierized basins, when a large area of ice is exposed, discharge increases rapidly following storms. Large precipitation events in winter and spring have little impact on runoff, yet in autumn storms produce significant runoff peaks.

More heat to raise snow to the melting temperature, and also delay the attainment of that temperature in spring, so the peak flows occur later at higher elevations. Irregular thickness of snowcover, in topographic hollows, and irregular melting patterns resulting from the effects of aspect and slope on radiation, with both direct exposure to the sun and shadow protection, lead to a protracted sustained snowmelt runoff period in high mountains rather than intense flooding. With increasing elevation, a larger proportion of total annual runoff occurs during summer months.

The impact of the existence of even a small area of ice-cover on the seasonal variation of runoff is marked. Snowmelt runoff is further retarded in glacier basins since glacial catchments extend to higher elevations than those which are solely nival. However, the major nival / glacial contrast results from the limited life of the snow 'reservoir', depleted during melt to disappear altogether, as compared with glacial ice which, available to release meltwater as heat supplied throughout the summer. Increasing percentage glacierization is associated with a larger proportion of catchment area at higher elevations, reducing winter runoff, and raising the percentage of annual total flow occurring between May and September in Swiss Alps and between April and November in the Himalaya and Karakoram mountains.

A characteristic feature of glacierized mountain rivers during the melting season is the pronounced diurnal variation of discharge. Diurnal fluctuations in the supply of energy available for melting snow and ice lead to a rhythmic diurnal flow regime, the hydrograph resembling saw-teeth, with daily peaks superimposed on a relatively steady background flow. In glacierized catchments the diurnal fluctuations are subdued early in the melt season and meltwater has to percolate through snow-pack; as the ablation season runoff from nival mountain basins varies from year to year as a direct result of precipitation fluctuations, and runoff cannot exceed precipitation. By contrast, in glacierized watersheds, year-to-year variations of runoff will be influenced additionally by storage remaining from one year to the next. Addition to and released from the mass of water stored as ice, annual total runoff may be either less or greater than the precipitation, and glacierized basins are therefore characterized by changes in snow and ice storage between years.

There is no direct relationship between years of high precipitation and high runoff. However, reduced variability of annual runoff in comparison with precipitation suggests a compensatory effect, in that precipitation is low, flow will be maintained by melt from glacier storage. Nevertheless, considerable differences in both quantity and timing of runoff exist between years. Changes in storage as ice are related to the climatic environment of the catchment and glacier over a time scale of several years. If, over several years, hydrometeorological conditions favour either growth or decline of glaciers and therefore of the volume of storage, then annual fluctuations of the quantity of runoff will be superimposed on an underlying trend determined by climatic shifting. Prediction and understanding of mountain runoff require measurement that will allow assessment of the effects of continued slow adjustment of glaciers to the fast changing state of the atmosphere.

In addition to storage as snow and ice, liquid water may be stored in glaciers and in marginal lakes against glaciers. An outstanding characteristic of glacial runoff is the occasional outburst flood created by the sudden drainage of large quantities of water stored in englacial, subglacial and ice-marginal sites. Even small water bodies or lakes can produce very large floods, because of the abrupt release of water. Some lakes may dump several times each year, others annually, and larger lakes every two to five years.

High mountain's hydrology is concerned with problems different from that of low elevation area. Many conventional hydrological prediction techniques cannot be used; as usual precipitation-runoff models are inappropriate. A unique set of physical hydrological processes is associated with snow and ice in mountains. Investigation of these processes requires information about the properties of snow and ice and the behaviour of glaciers in addition to an understanding of the components of the usual liquid water equation. Water cycle and climatic conditions are variable. As a result, the mass of glacier changes, and this leads to variations in water resources.

2.2 GLACIAL HYDROLOGICAL SYSTEM

The liquid water moves through the glacial system in several ways. The difficulty in conceptualizing complete-basin hydrology lies in the fact that there are seasonal dimensions involved (Young, 1982). Movement of the snowline on the glacier is slow and uniform; in contrast the snowline movement in the early ablation period on the non-glacier surface is much faster and non-uniform. The internal plumbing system of a mountain glacier consists of vertical pipes (moulins) connecting with a network of basal conduits joining to one or several major tunnels leading to portals (Figure 2.2). Whether basal tunnels are incised into bedrock or cut upwards into ice will affect whether flowing waters can access to bed areas by melting of channel margins as discharge increase diurnally or seasonally or by migrating with respect to fixed basal coordinates independent of glacier motion (Collins, 1987). In early spring and winter, conduits in ice will close, causing water to become stored by being sealed off in small cavities and channels, especially at the base of the glacier (Tangborn *et al.*, 1975). The other cause which tends to close basal passage ways during winter is due to snow over burden pressure. But when melting resumes in late spring, meltwater reaching the bed is unable to escape, water pressure is slowly raised as the summer advance and water spreads onto the bed, slowly expanding into linked cavity system.

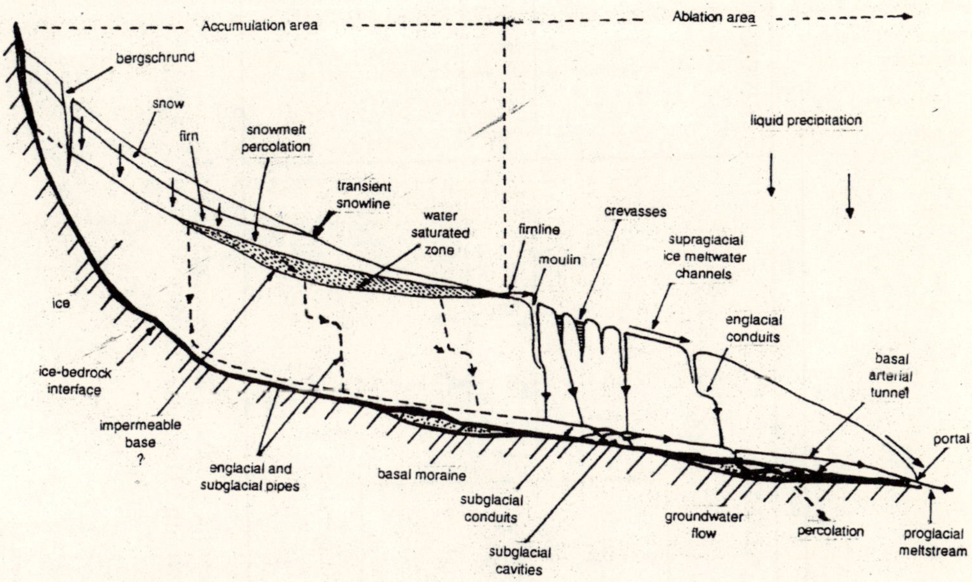

Fig. 2.2 The internal drainage system of a mountain glacier
Reprinted from Rothlisberger and Lang (1987)

Variations of atmospheric and cloud conditions cause glacier fluctuations. Nakawo and Hayakawa (1998) have shown in Figure 2.3 the process of glacier dynamics, which cause a delay in the water cycling, as well as the processes of water and heat exchange accompanied by thermal effects resulting from phase change of the water. As shown in the flow chart, glacier mass balance is intimately related to climate, river and glacier itself. The mass balance (water budget) is calculated as the sum of accumulation (amount of snow and ice added to a glacier: taken as positive) and ablation (amount of snow and ice lost from a glacier: taken as negative. Accumulation comes from solid precipitation and from sublimation (condensation) of water vapour, both of which are input directly from the atmosphere to the glacier surface. There can also be indirect accumulation, coming from avalanches, snow drifting and freezing of melt water and rain inside the glacier. This type of accumulation is very pronounced in the Himalaya.

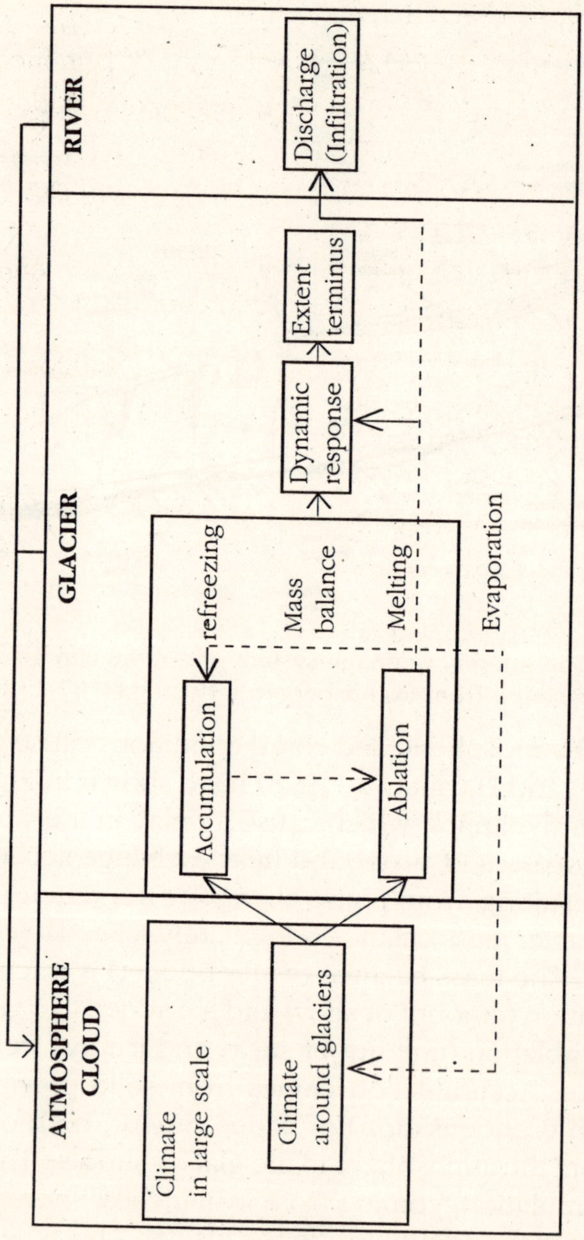

Fig. 2.3. Flow chart of glaciers related phenomena involving the atmosphere and hydrosphere. Reprinted from Nakawo and Hayakawa (1998)

Nakawo and Hayakawa (1998) have stated that the accumulation and ablation have annual cycles dependent on seasonal climates and the cumulative results of annual mass balance cause glacier variations. Since the glacier observations started in Europe and the U.S.A., it was originally thought that glaciers accumulate by snowfall in winter and that glacier mass change in summer is due to ablation only. Recent studies by Ageta and Higuchi (1984) have shown that, in Asia, the major part of both annual accumulation and ablation occurs simultaneously in summer, because little snow falls in winter due to the arid climate. For example, in the Himalaya and southern part of Tibetan Plateau, summer precipitation is high because of the Indian monsoon; it falls as snow at high altitudes.

2.3 MELT PROCESSES AT THE GLACIER SURFACE

The energy (Q_M) available for melt on a snow and ice surface of unit area and over a unit time consists of the following components:

$$Q_M = Q_{NR} + Q_S + Q_L + Q_P + Q_G$$

where:

Q_M = Heat used for melt

Q_{NR} = Net radiation

Q_s = Sensible heat transfer

Q_L = Latent heat of vaporization transfer

Q_P = Heat provided by precipitation

Q_G = Heat from heat conduction in snowpack

If all the components of energy budget are known, the melt rate (melted water equivalent of snow and ice per unit area and per unit time) can be computed:

$$M = Q_M / S$$

Where S is the latent heat of melting:

(1) Net Radiation (Q_{NR})

In basin with heterogeneous surface conditions, it is advisable to determine each component of net radiation separately (i.e. incoming and outgoing, short-wave and longwave radiation fluxes), to facilitate accurate interpolation and extrapolation.

Net radiation can be estimated from the folloiwng equation:

$$Q_{NR} = G(1-\alpha)+ \epsilon_a \ \sigma \ Ta_e^4 - \epsilon_e \ Te_e^4$$

Where:

G = Global radiation

α = albedo

ϵ_a and ϵ_e = emissivity of the atmosphere and of the surface respectively for snow a value of 0.98 – 0.99 is used.

σ = The Stefan – Boltzmann constant (5.67×1.0^{-8} J M^{-2} K^{-4} S^{-1})

Ta_e and Te_e = radiation temperatures of the lower atmosphere and of the ice surface respectively. If no direct measurements of the radiation fluxes are available, the shortwave net radiation, $G(1 + \sigma)$, can be estimated from sunshine duration records or from cloud observations, by taking into consideration latitude, time of year, exposure and slope. Longwave net radiation can also be estimated from empirical relationships based upon cloudiness.

(2) Sensible heat (Q_S) and latent heat (Q_L)

The contribution of the turbulent fluxes of sensible and latent heat to the energy balance is very variable and depends mainly on the local meteorological condition. Direct measurements are very difficult to perform. The turbulent fluxes are usually derived from temperature, relative humidity and wind speed measurements in the lowest boundary layer. The turbulent heat fluxes were calculated using several approaches.

(a) *Profile Method*

The turbulent fluxes are derived from measurements at 4 levels which were located at Z = 0.3 m, 0.8 m, 1.8 m and 3.2 m above the snow surface.

The profile can be described by the Businger-Dyer equations (Still, 1998 and Holtslag and Van Walder, 1983).

Sensible heat flux

$$H = -PC_P\theta$$

Latent heat flux

$$LE = -PL_V\,q$$

with

$$U = K(u_2 - u_1)\{\ln(Z_2/Z_1) - \psi_m(Z_2/L) + (m(Z_1/L)\}^{-1}$$

$$\theta = K(\theta_2 - \theta_1)\{\ln(Z_2/Z_1) - \psi_h(Z_2/L) + (h\,Z_1/L)\}^{-1}$$

$$q = K(q_2 - q_1)\{\ln(Z_2/Z_1) - \psi_e(Z_2/L) + (e(Z_1/L)\}^{-1}$$

The Obukhov length is defined by :

$$L = \theta\,u^2/K_g$$

U and Q are the wind speed, the potential temperature and the specific humidity of the air as they are measured at different levels Z, K is the Von Karman constant, taken as $K = 0.4$, g is the acceleration due to gravity, C_p specific heat of air and L_v latent heat of vaporization of water.

The ψ- functions ψ_m for mass, ψ_h for heat and ψ_w for water vapour are given by Paulson (1970):

ψ for $L < 0$ (unstable)

$$\psi_m = 2\ln(1 + x/2) + \ln(1 + x^2/2) - 2\tan^{-1}(x) + \pi/2$$

$$\psi_h = \psi_w = 2\ln(1 + y/2) \text{ with}$$

$$x = (1 - 15\,Z/L)^{1/4},\ y = (1 - 9\,Z/L)^{1/2}$$

for $L > 0$ (stable)

$$\psi_m = \psi_h = \psi_w = -6\,Z/L$$

The calculations were performed interactively between any 2 measurement levels Z_1 and Z_2. The roughness length Z_0 is calculated using the following equation:

$$Z_0 = K(q_2 - q_1)\{\ln(Z_2/Z_1) - \psi_h(Z_2/L)\}^{-1}$$

in near neutral conditions where the ψ functions can be neglected. For each neutral run (where $Z/L < 0.05$), Z_0 was calculated as the height Z_1 where $U_1 = 0$. The roughness lengths obtained in this manner were averaged over a da.

(b) *Eddy Correlation Method*

The measurement of wind speed by an acoustic anemometer enables the sensible heat flux to be determined directly.

$$H = PCP_w \text{'}T\text{'}$$

where:

$W\text{'}T\text{'}$ = vertical kinematics eddy heat flux

The roughness length Z_0 can also be calculated from the sonic measurements for near neutral condition:

$$Z_0 = Z/(K_u(Z)/u)$$

using $u^2 = u\,\text{'}w\text{'}$ exp from sonic measurements

(c) *Bulk Method*

The turbulent heat fluxes are often approximated over melting snow with only one measurement of temperature, relative humidity and wind speed above the snow surface. This method is widely used because of its simplicity. Over melting snow at height Z_0 the surface temperature is taken as 0°C and at height Z_0 the saturation vapour pressure as 6.1 mbar. With this assumption the Buinger-Dyer relations can be used to calculate in fluxes.

2.3.1 Computation of Melt Rate

(a) Energy Balance Approach

According to Upadhyay (1995) the net change in energy storage of a glacier or snow cover (ΔQ) is expressed as:

$$\Delta Q = Q_N + Q_C + Q_E + Q_G + Q_M + Q_R \qquad ...(1)$$

Each term has the unit of Wm^{-2} or langley/day and is +ve if it adds to the energy storage of glacier mass as shown in Figure 2.4.

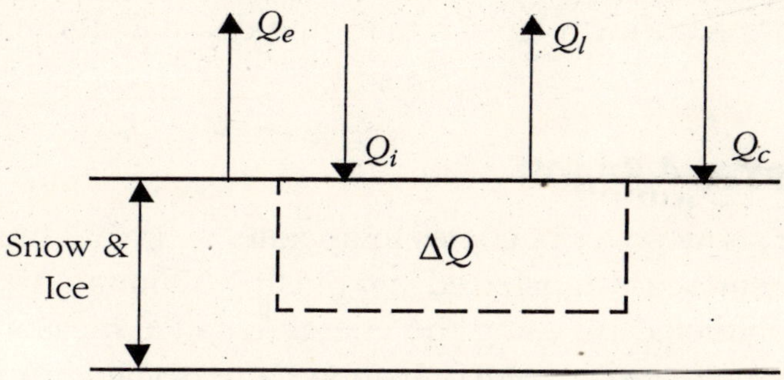

Fig. 2.4. Energy storage of glacier ice mass.
Reprinted from Upadhayay (1995)

Q_n = Net radiation received by the glacier surface

Q_c = Heat transfer at ice-air interface by conduction (sensible heat, +ve of air temperature (T_a) > snow surface temperature (T_s) and –ve if ($T_a < T_s$).

Q_e = Latent heat flux due to evaporation / sublimation(–ve) or condensation/sublimation (+ve)

Q_g = Heat exchange at snow-ground interface (+ve but very little).

Q_m = Advected heat flow (mass transfer of vapour or meltwater or rain water) (+ve or –ve)

Thus, net radiation (Q_n) for given time interval Δt is:

$$Q_n = Q_i (1 - \alpha) + Q_a - \epsilon \sigma T_s^4 \Delta t$$

where Q_i = incident solar radiation (direct and diffused)

α = albedo of surface (fraction)

Q_a = incident longwave radiation emitted by atmosphere.

σ = Stefan Boltzman constant (5.67×10^{-8} w m^{-2} k^{-4} or 1.355 $\times 10^{-12}$ ly k^{-4} S^{-1})

T_s = snow surface temperature (K)

Upadhyay (1994) has suggested for computing the physical/meteorological parameters, the snow covered catchment may first be divided into various altitudinal zones. Then the temperature of each zone may be computed for preparing a lapse rate for the entire altitudinal elevations.

(b) Melt Energy and Runoff

Once the entire snow cover becomes temperate, nearly all the energy flux through the surface is transferred into melting (Ohmura, 1981). The runoff can be predicted by using the energy balance measurements. Surprisingly, the degree-day approach allows a reasonable runoff prediction, despite the fact, that it mainly describes the sensible heat flux component of energy balance.

2.3.2 Energy Balance Studies in Alpine Environments

Net radiation dominates the energy balance in most studies, but its absolute value varies considerably. The sensible heat flux ranges from negligible to equal in importance to net radiation. The latent heat flux is negative and relatively small in most areas. It is frequently opposite in sign but of similar magnitude as the sensible heat flux, hence the net turbulent heat flux is very small. This might be the reason why energy balance approach leads to good averaged results for melt energy determination.

2.3.3 Snow Melt Modeling

In glacierized environments net radiation generally dominates the energy balance and is the major energy source for snow-melt. The contribution of the turbulent heat fluxes to snow-melt can also be of major importance in some climatic conditions. Their magnitude is extremely variable and depends largely on the wind speed. The results of various studies show that runoff prediction from an alpine snow cover is possible using an energy balance approach with good accuracy. While average runoff can be modeled with relatively simple models, the daily variations are more sensitive and demand the exact knowledge of the energy balance.

2.3.4 Degree-Day Method

A degree-day mass balance model has been developed to estimate the effect of climate changes due to greenhouse warming on glaciers in the Nordic countries. The model computes the mass balance of a glacier from information about the temperature and precipitation at meteorological station in the vicinity of the glacier (Johannesson T. *et. al.*, 1993). The following are the advantages in degree-day model over the energy-balance model.

1. The formulation of the mass balance in terms of temperature and precipitation is advantageous for simulations of the response of glacier mass balance to climate changes. Scenarios for future changes in cloudiness, cloud height, wind speed, relative humidity and other variables which are often required in energy-balance models (Oerlemans, 1992; Braithwaite and Olesen, 1990) are in general not available.
2. Degree-day coefficient shows a high degree of stability with time and across a wide range of different climatic regimes.

2.4 MECHANICS OF WATER PASSAGES

Meltwater produced at the surface is the most important source. The water moves through the ice by small and large passages, such as moulins and glacier tunnels. Ice is deformable, so that passages in it can expand and contract significantly in response to increase and decrease in water pressure on their walls relative to the ambient pressure in the enclosing ice as pointed out by Glen (1954). The behaviour of water passages is primarily the result of three basic characteristics:

1 The capacity of passages continually adjusts, to increase and decrease in the supply of meltwater.

2. In a steady state the pressure, and the movement, of water in them is governed primarily by the ambient pressure in the enclosing ice and by the rate of melting of the passage walls.
3. Due to the differential growth of the larger passages, the network of passage tends in time to become arborescent, with tributaries

joining into ever-larger trunk passage, like a three dimensional river network.

These three characteristics explain many features of glacier hydraulics.

2.5 CHARACTERISTICS OF GLACIER RUNOFF

The controlling factors on the glacier discharge are the heat fluxes and surface energy balance. Precipitation (solid or liquid) generally has a negative influence on glacier runoff because incoming solar radiation is reduced because of cloud cover and the new snow has a high albedo. Variations in glacier runoff follow approximately the reverse pattern of a rain-dominated runoff regime. As a result, it is possible to observe a compensation effect in mountain river basins where the upper parts of the basins experience a meltwater runoff regime (nival and glacier regime) and at the same time in the lower parts of the basins runoff is dominated by rainfall. In glaciated areas, much of the precipitation falls in solid form throughout the year, so that it contributes to mass storage rather than directly to runoff. For example in the Alps, a minimum variation in annual runoff is observed from river basins with 30-40 % glacier cover (Kasser, 1959). Figure 2.5 shows a plot of coefficient of variation (CV) of runoff in relation to the percentage of glacier area for some alpine river basins. The amount of precipitation in the form of snow and ice has an inverse affect on the amount of runoff. Precipitation strongly effects the reflectivity; new snow is highly reflective so that it absorbs less heat and therefore melts more slowly. Old snow, firn, and glacier ice have a low reflectivity. Thus greater the precipitation in the form of snow, the longer the glacier is covered by a highly reflective material, and the less the runoff. A decreased amount of snowfall leads to low-reflectivity surface being exposed longer than usual producing greater melt and increased runoff. Thus runoff from glaciers are naturally regulated, producing increased runoff at a time of high temperature or deficient precipitation, and storing water when the need for it is lower.

2.6 DIURNAL AND ANNUAL CYCLE

The diurnal and annual cycles of glacier runoff are primarily controlled by incoming solar radiation and temperature variations. Figure 2.6 shows the diurnal discharge pattern of a small Himalayan glacier. Runoff from the basin (24 km^2, 45% glacier cover) is characterized by a considerable base flow upon which is superimposed a diurnal cycle of runoff. Similar example from the Austrian Alps is given in Figure 2.7 of the seasonal development of the glacial drainage system as reflected in the diurnal variations in runoff of Austrian Alps.

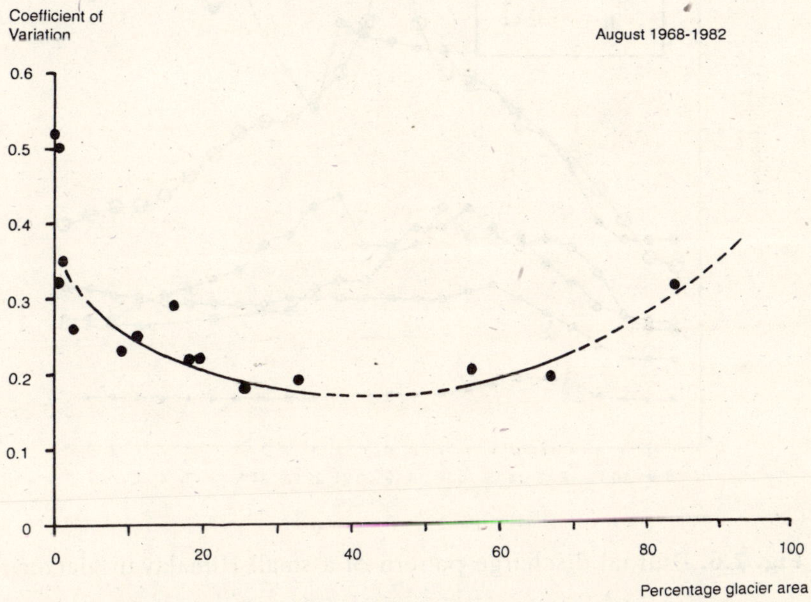

Fig. 2.5. Coefficient of variation (CV) of runoff in relation to the percentage of glacier area for alpine river basins. Reprinted from Rothlisberger and Lang (1987)

The studies of Lang, 1973; Lang *et al.*, 1979; Collins, 1982 on hydrograph analysis and tracer experiment give a better picture of the glacial drainage system. From such studies it appears that the subglacial flow consists of several different components: groundwater, water filled cavities, and the firn water acquifer which is fed by summer melt processes in the accumulation area of the glacier, while the superimposed diurnal cycles of runoff consist of rapidly draining components of that day's meltwater. The theory of the main subglacial drainage conduits as the result of water pressure was developed by Rothlisperger (1972).

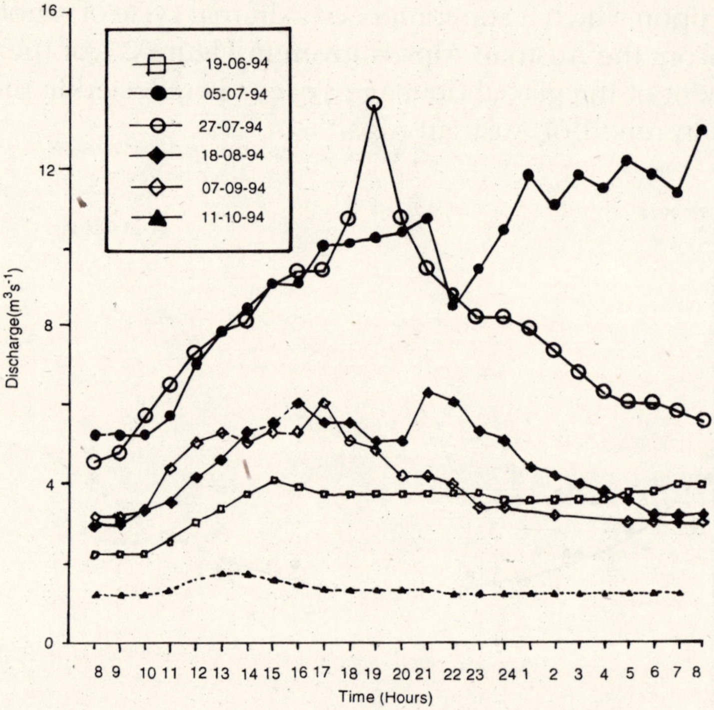

Fig. 2.6. Diurnal discharge pattern of a small Himalayan glacier

2.6.1 Water Storage

Liquid water is effectively stored in several locations within an alpine glacier (Collins, 1982), such as surface snow, pools on ice surface, firn, englacial pockets, subglacial cavities, the moulin-conduit network and basal moraine, while in transit through various hydrological pathways of the internal drainage system.

Water storage in glaciers also fluctuates over periods of several days with hydrometeorological conditions, resulting in melt-stream discharge having a diurnal peak, superimposed on a background flow, the volume of which varies move slowly. Only a portion of the meltwater produced each day reaches the portal on the same day. Runoff of the remainder is delayed until succeeding days, adding to later daily melt contributions (Martinec, 1970).

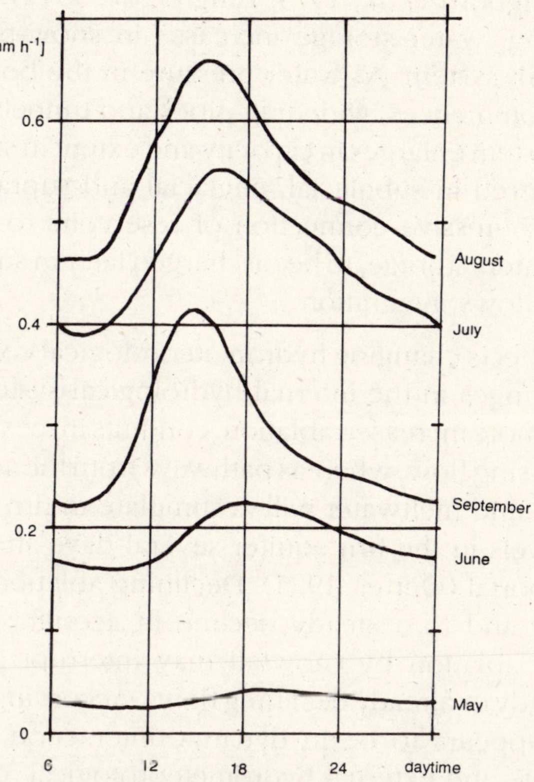

Fig. 2.7. The progressive development of the Swiss Alps glacier drainage system during the ablation period is clearly reflected in the changing characteristics of the diurnal variations in runoff from month to month (illustrated by changes in the time of daily maximum flow, the gradients of rising and falling limbs of the hydrograph). Reprinted from Rothlisberger and Lang (1987).

The internal conduit system to allow meltwater to runoff rapidly depends on the extent to which the tendency for channels to open by frictional melting is offset by closure resulting from the ice overburden pressure (Rothlisberger, 1972). Opening of the passages follows from increasing discharge, and closure from declining flow. Water pressure in conduits thus varies at both seasonal and diurnal time scales. It is generally accepted that in winter, and early spring, when there is little ablation, conduits in ice will close, causing water to become stored by being scaled off in small cavities and channels, especially at the base of the glacier (Tangborn *et al.*, 1975; Lang *et. al.*, 1977)). As ablation resumes, in spring, water storage increases in snow-pack, firn and in moulin - conduit system. As water pressure in the body of the glacier rises, so flow commences, widening pipes and tunnels. As the internal hydrological system enlarges in capacity and extent in summer, drainage of the water stored in subglacial, englacial and supraglacial locations ensues with progressive connection of reservoirs to runoff. Thus, in spring, water enters storage, to be discharged later in summer as conduit development allows integration.

Storage reflects changing hydrometeorological conditions and corresponding changes in the internal hydrological system. If favourable conditions promote increased ablation, conduits in ice will rapidly widen allowing increasing flow, whereas pathways from the accumulation area do not expand and meltwater will accumulate in firn leading to maximum water levels in the firn aquifer several days after maximum discharge at the portal (Oerter, 1984). Declining ablation leads rapidly to declining flow and to a steady decline in acquifer storage. Sudden termination of ablation by snowfall may interrupt periods of either increasing, steady or already declining flows. According to Collins (1982) stored water appears to be in dynamic interaction with changes in conduit capacity and external hydrometeorological conditions. Short-term periodic variations in storage appear to be more important in the formation of runoff than seasonal trend.

2.7 MELTWATER SYSTEMS

Meltwater may exist either within a series of discrete and segregated hydraulic systems in each individual glacial sub-environments or as a few large systems that interconnect with all or most sub-environments. The development of glacial hydraulic systems within any ice mass is dependent upon: the type, morphology and topographic setting of the ice mass; the mass balance; basal conditions; the seasons and climatic conditions present and past; and the amount and type of debris cover and content. Meltwater systems may be in the form of: (a) thin sheet flow; (b) channel flow; (c) linked cavity flow; or (d) an integration of above forms. Meltwater systems may be seasonal or of shorter duration, and may exhibit normal and catastrophic fluctuations. The frequency and magnitude of system discharge fluctuations may vary widely from season to season, on different part of same ice mass and between neighbouring ice masses. The evolution of different systems may also vary enormously in time and space. Drainage system in temperate glaciers during the melt season, however, will consist of three distinct parts, which for convenience may be termed as supraglacial, englacial and subglacial. Figure 2.8 illustrates various channels of the drainage system in a Himalayan glacier. Supraglacial and englacial hydraulic systems exhibit considerable fragility, being easily susceptible to rapid change whereas established subglacial systems, most resistant to alteration under ice conditions, are dramatically transformed as in a surge or rapid ice retreat.

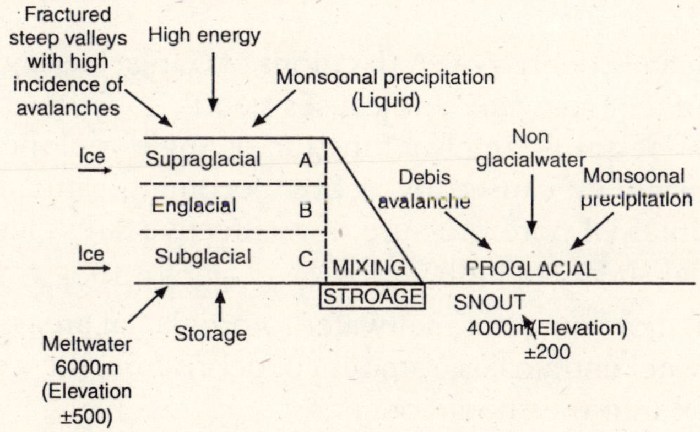

Fig. 2.8. Various channels of the drainage system in typical Himalayan glacier

2.7.1 Supraglacial Hydraulic Systems

The supraglacial hydraulic systems develop an arborescent pattern largely in accordance with subaerial fluid dynamics. As overall discharge and/

or gradient increases, the hydraulic system tends to evolve from disconnected flow to an arborescent drainage pattern of connected channels. Supraglacial lakes may develop, especially in summer months, and form part of the drainage system over a large area of the ice surface. Drainage on the surface may also be interrupted by meltwater disappearing into englacial and subglacial systems via moulins. Depending upon the mass activity, vertical ice thickness, state of internal stress and other local factors, connections may or may not be made between supraglacial and englacial and/or subglacial hydraulic systems.

In valley glaciers confined within a bedrock channel with high surrounding valley sides where a distinct transverse profile (the result of differential flow velocity) develops, the profile leads surface meltwater to flow towards valley sides. Lateral streams may then flow either along the edge, often disappearing within lateral moraines or submarginally beneath the ice. Where glaciers are extensively covered by morainial debris over large area of surface, as in the Karakoram-Himalaya mountain systems, supraglacial stream flow is interrupted by moraines and much of the surface meltwater disappears into the debris. Supraglacial meltwater in the Himalayan glaciers influences several sedimentalogical processes:

1. The heavily debris covered regions of Ganga basin glaciers soaks up surface meltwater and causes debris flows. Such mass transport of debris is largely a function of angle of slope, gravity and the instability caused by debris becoming saturated, reaching high internal pore pressure, low effective stress levels and low internal shear strength.
2. As a major source of meltwater to proglacial areas, supraglacial meltwater entrain large amount of debris from the ice surface and push it into the snout zone.
3. The total hydraulic system of a glacier has been likened to an enormous "plumbing system" in which the supraglacial system acts as one of the sources whereby debris is flushed out through the system. Figure 2.9 shows a very well developed supraglacial channel on the Chhota-Shigri glacier. Lahul-Spiti valley, Himachal Pradesh, India.

Fig. 2.9. Well developed supraglacial channel, Chhota Shigri glacier, Himachal Pradesh

2.7.2 Englacial Hydraulic Systems

The englacial hydraulic systems may be viewed as a complex architecture of interconnected cavities and conduits within a three dimensional gallery system. Where ice is temperate and fast moving, englacial systems are likely to be temporary and largely autonomous. Under slow moving and temperate conditions, a well developed englacial system that may interconnect to the subglacial system may develop. This network starts near the surface with capillary tubes, which then join downward to form increasingly larger channels. At intervals, crevasses or moulins provide much larger direct connections between the surface and deeper parts of this system. Some of the larger passages will reach all the way to the surface as moulins to receive the supraglacial runoff, but most divide and re-divide upwards to drain the water percolating downwards. Those englacial channels which show wide variations in water pressure are believed to have direct connection with subglacial hydraulic system.

In general, englacial hydraulic systems are best developed in ablation areas of ice masses. The style, geometry and degree of development of any englacial hydraulic system are function of : the meltwater discharge entering the system; the depth, the width and number of crevasses to be exploited; the internal distribution and rheological state of the ice mass; the conditions at the bed of the ice mass in terms of sliding velocity and subglacial hydraulic system type and level of development; and the overall stability of the ice mass in the long-terms.

For an englacial conduit in order to remain open and stable under active ice conditions and after the crevasse seals, it is necessary for those forces maintaining the opening of the conduit to equal those attempting to close it. Englacial conduits remain open as a result of thermal energy generated by: (a) meltwater turbulence; (b) frictional energy due to transported debris within meltwater causing sidewall abrasion; and (c) the diffusion of direct thermal energy from the passage of warm meltwater. The factors leading to conduit closing are those associated with the intrinsic internal deformation of the confining ice mas attempting to seal the conduit space. Also, when englacial meltwater discharge wanes for whatever reason , debris transported in these channels will settle and possibly clog the system.

2.7.3 Subglacial Hydraulic Systems

The imprint of subglacial hydraulic systems is unquestionably one of the most important features of any glacial landscape. Lying as it does at the interface between the glacier and its bed, the subglacial system is capable of moving vast quantities of sediment, of cutting down into bedrock and already existing deposited sediment, and thus imparting a critical influence upon glaciodynamics that ultimately dictate the 'effect' of glaciation upon any land. The subglacial part of the drainage system will be more complex because of the presence of a sediment load and the influence of bed rock topography. Recent dye tracing studies carried out on some glaciers in Switzerland, indicate that the subglacial drainage system contains two principal components: a distributed system in which flow velocities are of the order of 0.5 m/s, and a channelized system in which flow velocities are in the range of 0.5–1.9 m/s. During the melt season the channelized system expands headwards at the expense of the distributed system so that by the end of the season it occupies the lowermost 3-3.5 km of the 4 km long glacier. Headwards growth of the subglacial drainage channels closely follows the upglacier retreat of the transient summer snowline on the glacier surface. In areas where the ice is in direct contact with a horizontal or downglacier-sloping, nondeformable, undulating bed, a tortuous linked-cavity system will be present, transected by and in hydraulic contact with one to several larger straighter conduits that are likely to be broad and low. It is also clear that the positions of some or all of these conduits may change seasonally and , from time to time, permanently (Hantz and Lliboutry, 1983). The latter changes are apparently due to secular changes in thickness and surface slope of the glacier that are sufficient to change the positions of the equipotential contours. In general, however, the average trend of conduits should be normal to the equipotential contours on the bed. In areas of deformable bed, the subglacial drainage may be either through a permeable layer or till, through a large number of quite small subglacial passages, or through englacial conduits. This drainage configuration suggests that two-component model adequately describes the subglacial transport of meltwaters.

Therefore, the total (or bulk) discharge consists of two components. The first, quickflow, is derived largely from icemelt, which passes rapidly through the glacier in ice-walled channels (englacial flow), and has only limited contact with sediments. Hence, quickflow maintains the composition of ice melt and is extremely dilute. The second, delayed flow, is derived — largely from snowmelt, and drains relatively slowly through the distributed system at the ice-bedrock interface. Before draining from the glacier snout, the two components mix in the major arterial channels at the glacier bed.

In the mountainous regions of Karakoram and Himalaya where fresh water and/or non-polluted water is in limited supply, subglacial meltwater has been found in recent years, to be potential and tappable long-term reliable source of water.

2.8 HYDRAULIC SYSTEMS AND DEFORMABLE BEDS

Hitherto hydraulic systems have ben largely considered with rigid beds (H-bed types). However, increasing evidence reveals that large portions of many glacier beds today and in the past were probably composed of soft, unconsolidated beds (M-or Q-bed types) termed deformable beds.

Where deformable sediment, saturated by meltwater, exists beneath ice masses, deformation will occur when the applied shear stress imparted by the overlying active ice overcomes the internal shear strength of the sediment. Perhaps, the greatest controlling variable influencing this process is the intrinsic viscosity of the sediment, since this parameter reflects the level of consolidation, grain size and porosity of the sediment. If it is assumed that the sediment at the ice-bed interface is deforming, then it is generally thought that little or no free meltwater derived from solely subglacial sources, would exist at the upper interface between the mobile sediments and glacier sole. However, where meltwater from external sources reaches the glacier bed (as in the Garhwal, Himalaya from monsoonal rainfall), a slightly different set of circumstances may occur. In the first instance, extraneous water cannot be totally absorbed by the already saturated sediment and thus moves along the above interface. The absorbed meltwater will aid in reducing sediment viscoisity, further allowing the sediment to deform at a greater rate. The remaining unabsorbed water, however, must find a pathway along the upper interface or within the sediment itself in the form of pipe flow.

2.9 LONG-TERM VARIATIONS

The long-term variation in glacier runoff consists of two components:

1. There exist a direct relationship between climatic variations and accumulation/ablation processes, water storage and runoff.

2. Himalayan glaciers have been receding throughout the period AD 1850 to 1960 (Mayewski *et. al.*, 1980). The shrinkage is more visible in the snout sections of glaciers. Since these are the zones of highest melt rate, the progressive disappearance of these zones has resulted in a corresponding loss of potential for meltwater yield. Figure 2.10 shows the changes that occurred at snout between 1935 and 1971 in the Gangotri glacier, Garhwal Himalaya. Vohra (1989) has calculated the recent recession of the Gangotri glacier. He has estimated one million m^3 ice loss per year from the Gangotri glacier. The ice loss from its tributary glaciers and surrounding regions would be very substantial and is to be added to this figure. Current climatic models indicate a tendency towards lower precipitation in the middle latitudes which coupled with increasing temperatures, would result in a period of pronounced glacier retreat and thus a marked loss of meltwater runoff potential. Initial attempts to estimate the reactions of mid-latitude glacier mass balance to predicted climate changes have been reported by Kuhn (1985).

2.10 HYDROLOGIC MODEL FOR KARAKORAM MOUNTAINS, PAKISTAN

The upper Indus Basin which has an area of 960×10^3 km^2 consist of a series of mountain ranges of extreme ruggedness and very high elevations. Figure 2.11 shows the drainage map of Indus basin with Batura glacier. The ranges run approximately west-east in arc from the Hindu Kush and Pamirs in the west and north to the main Himalayan chain to the south and east. The latter culminates in the Nanga Parbat massif, which forms a barrier, to the northward movement of monsoon.

The glaciers of the Karakoram cover an area of about 16300 km^2. The dominant source of precipitation is depression coming from the west popularly known as westerlies and this source provides the snow for the glacier system of the Karakoram. A hydrologic model proposed by Young and Hewitt (1990) for Karakoram describes that most of the precipitation in this region is derived from westerly depression. In July

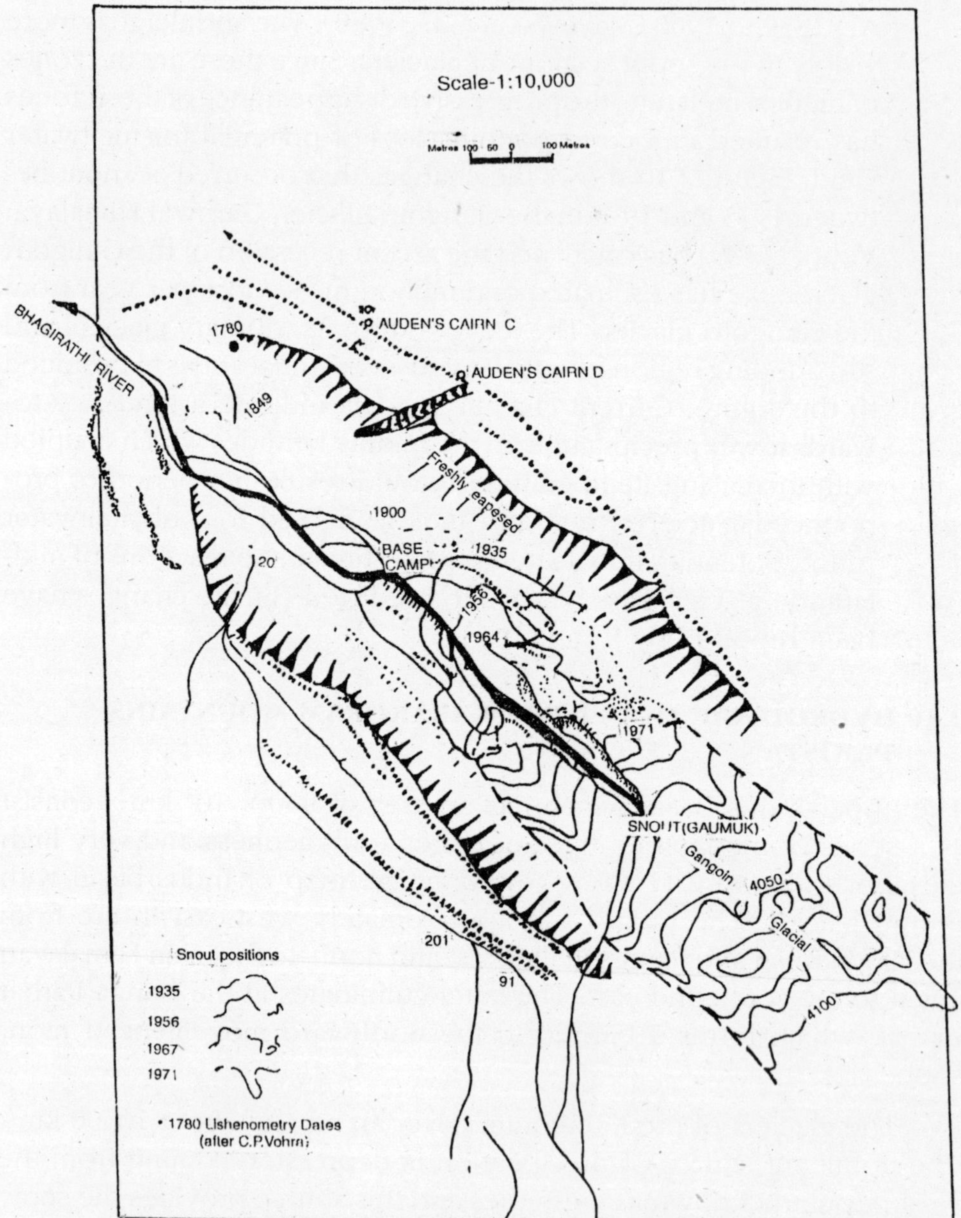

Fig. 2.10. Recessional trends of the Gangotri glacier, Garhwal Himalaya. Reprinted from Vohra (1989)

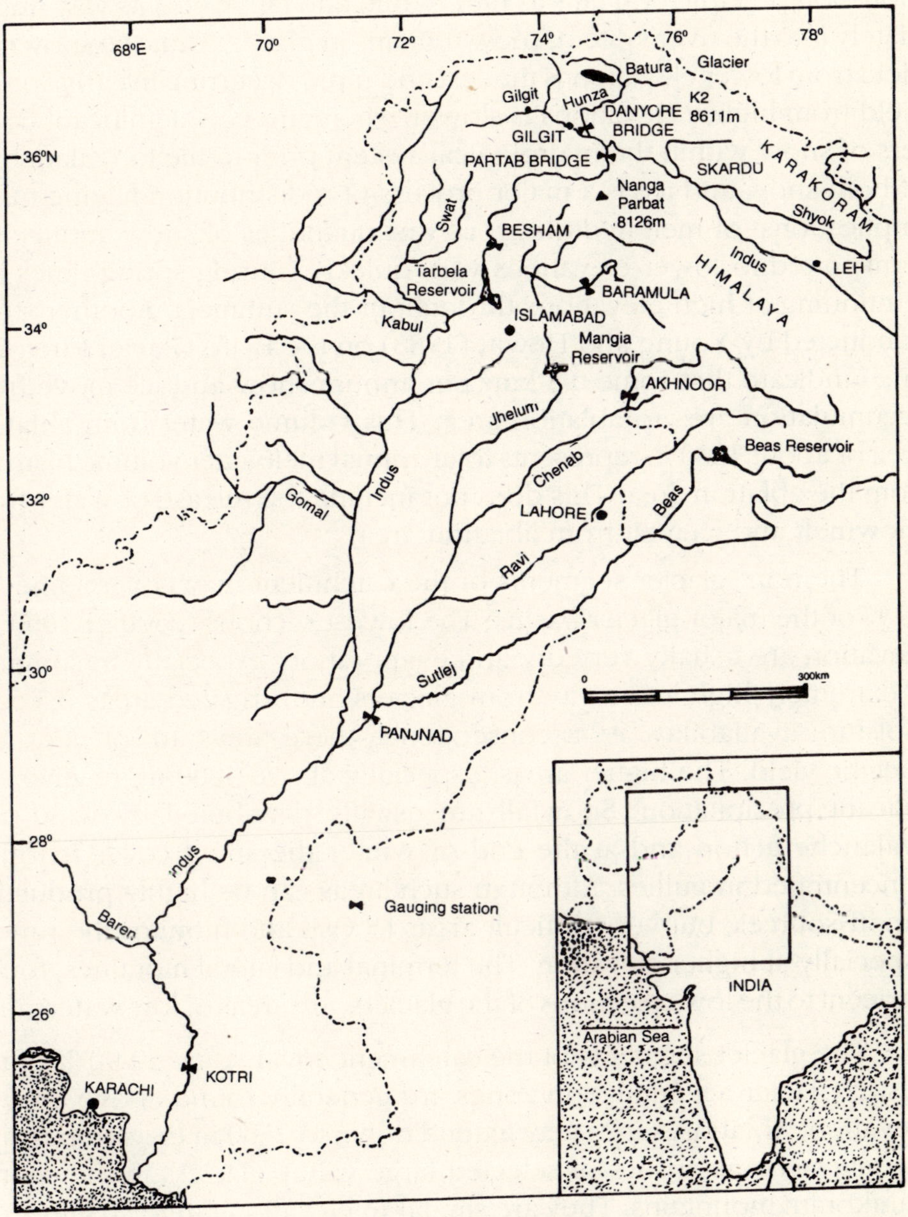

Fig. 2.11. Map of the Indus basin showing Batura glacier in the Karakoram mountains.

and August temperatures above freezing prevail up to 5000 m, hence the monsoon precipitation falls as rain. While rain fall makes its way immediately into the river system, snow remains in places. Thus the snowmelt yield from lower elevation is modest and rapid occurring in early spring, yield from higher elevations is slower. There are very significant transfers of snow within the hydrological system prior to melt. Avalanching of both snow and ice is a major process of redistribution having major implications of melt hydrology, as vast quantities of snow and ice are transported to lower elevations very quickly (during spring times but continuing at high elevations throughout the summer). Another study conducted by Young and Hewitt (1988) on the Biafo Glacier (area 267 km^2) indicate that some 0.5 km^2 per annum snow and ice move from accumulation area to ablation area. This volume water from ablation area of about 3000 m represents total annual meltwater volume from ice from the ablation area. This does not include the release of water from snowmelt and icemelt from ablation area.

The non -glacier segments of the catchment may cover some 10-50 % of the major glacier basins. The lowest sections (2500 or 3000 m) elevation are usually very dry and evaporation, especially in summer, is extremely high and away from glaciers and irrigated areas exceeds moisture availability. As a consequence these areas are of very low specific yield. The higher areas, especially above 4000 m, receive significant precipitation. Snowfall are usually distributed by wind and avalanche action and at the end of winter the snow cover is found concentrated in gullies. Although such areas can be highly productive runoff sources, but very difficult areas to visit and monitor the runoff, especially at higher elevation. The terminal and lateral moraines, found adjacent to the lower sections of the glaciers, are areas of low water yield.

The glacier segments of the catchment cover some 50-90 % of the basin. Glacier accumulation zones are generally found above 4800 m elevation. Ablation zones may extend down to 2500 m elevation. Table 2.1 gives characteristics of selected large valley glaciers in the central Karakoram mountains. They are similar in elevation range, in elevation of the equilibrium lines and in accumulation/ablation ratios. The major characteristics of their meltwater productivity are thus similar (Young and Hewitt, 1993). In years of heavy winter snowfall and/or in cool summers, the glaciers grow in size, storing water from one year to next. The converse holds in years of light snowfall and/or warm summers. Glaciers therefore perform the important function of regulating streamflow; probably some 70 to 80 % of the flow in the Indus river tributaries derived from glaciers in Karakoram and Nanga Parbat ranges. Temporal variability is an outstanding feature of the Upper Indus Basin. Figure 2.12 illustrates monthly discharges for the Indus at Beṣham, the Kabul at Warsak and the Hunza at Dainyor. Summer snowmelt with glacier ice melt produce the very high peaks of mid and late summer.

Table 2.1 Characteristics of selected large valley glaciers in the Central Karakoram.

	Biafo	*Pasu*	*Bualtar*		*Barpu*	*Rakhiot*
				Miar	*Sumaiyarbar*	
Basin area(km²)	853	104		← 414 →		102
Area of main glacier (km²)	536	62	80	53	65	32
Glacier area within basin (km²)	544	62	85	126		45
% of glacier cover	64	60		← 55 →		44
Maximum basin elevation (m)	7286	7683	7275	6824	7460	7900
Glacier terminus elevation (m)	3110	2769	2439	2835	3350	3200
Icefalls on glacier	None	Yes at 5000 m	Yes at 5000 m	Yes at 5000m	None on main stem	Yes
Debris cover	Extensive	Very little	Extensive Associated with landslides	Very little	Extensive	Extensive
Access to accumulation area	Easy	Very difficult	Very difficult	Very difficult	Very difficult	Very difficult

[1] The Bualtar/Barpu is one system. The Miar and Sumaiyarbar are the major components of the Barpu Glacier.

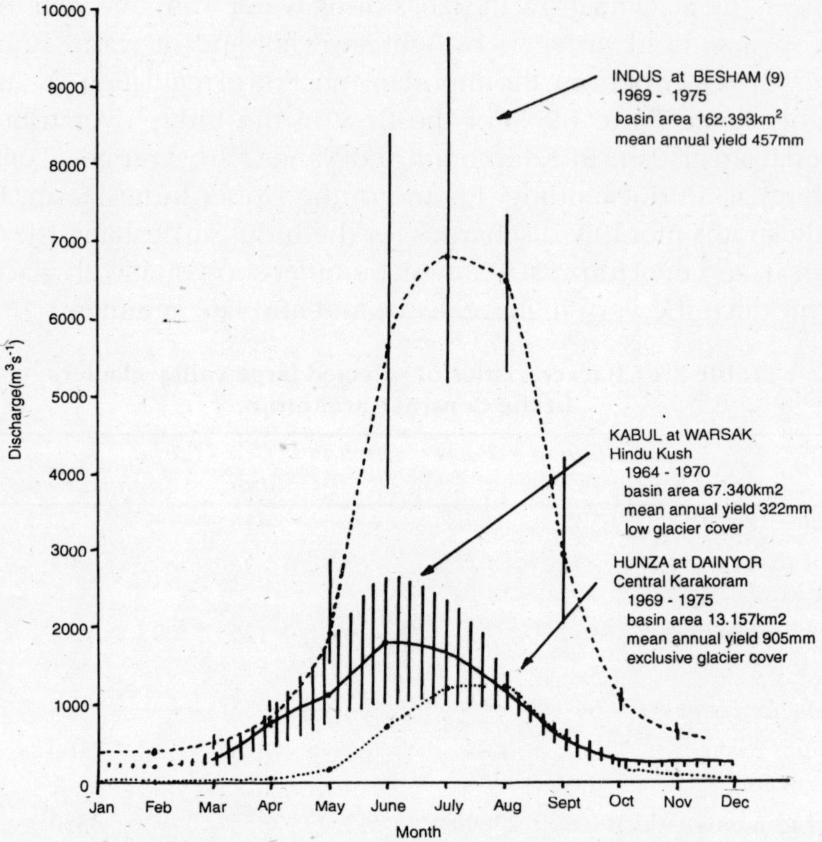

Fig. 2.12. Maximum, mean and minimum monthly discharge for three Indus basin rivers. Reprinted from Young and Hewitt (1990)

2.10.1 Hydrological Processes

There are seasonal variations in discharge of the Indus and its principal tributaries, the Jehlum, Chenab, Ravi and Sutlej. The headwater of the Indus and its tributaries are predominantly fed by glacier icemelt, spring snowmelt and little summer monsoon rains. Winter snowfall increases with both elevation and latitude westwards, whereas monsoon rains decreases to the north-west and behind mountain barriers reaching a maximum at about 2000 m.a.s.l (Barry, 1992). The quantity and timing of discharge in Karakoram rivers with snow and ice-melt components of runoff depend therefore on amount, incidence and form (Liquid or solid) of precipitation, and on thermal regime which determines the patterns of melting in spring of the amount of winter snowpack available and of the melting of perennial ice throughout summer in those basins which are glacierized.

A concpetual runoff model for the upper Indus Basin has been developed by Collins (1994) based on his field studies during which he has carried out the measurements of climatic variables, such as incoming global radiation, air temperature, lapse rate and precipitation throughout the ablation period with measurements of glaciological characteristics such as mass balance, ablation of ice and accumulation of snow. Seasonal fluctuations in components of the basin in hydrological cycle under various conditions in Karakoram mountain catchment are shown schematically in Figure 2.13. Water equivalent held in the winter snow pack will be greater and remain longer in spring at higher elevations. Both the quantities of water stored and the elevations at which the storage occurs affect the pattern of melt and the timing and the volume of release of meltwater from snow.

The pattern of runoff derived from the ablation of ice depends on the rate of removal of the winter snow cover. Essentially the rise of the runoff hydrograph in spring is controlled by the rate at which transient snowline rises, before decline takes over accompanying the reduction in radiation. Cloud cover cuts off the heat energy and the resulting ice-melt will be reduced during this period.

Batura glacier is located at about 36° N in the Karakoram mountains as shown in Figure 2.11. Meltwaters draining from the snout of the 60 km long glacier enters the Hunza river about 1 km downstrem from the glacier snout. Discharge, rainfall and daily mean air temperature of Batura glacier measured by Collins (1994) during the ablation season of 1990 (Figure 2.14) have characteristics which resemble those measured in the glacierized basins in Swiss Alps (Collins, 1989). Discharge started to rise in early May, although daily mean air temperatures at the terminus had reached 10°C by mid-April. Most of the glaciers remained covered

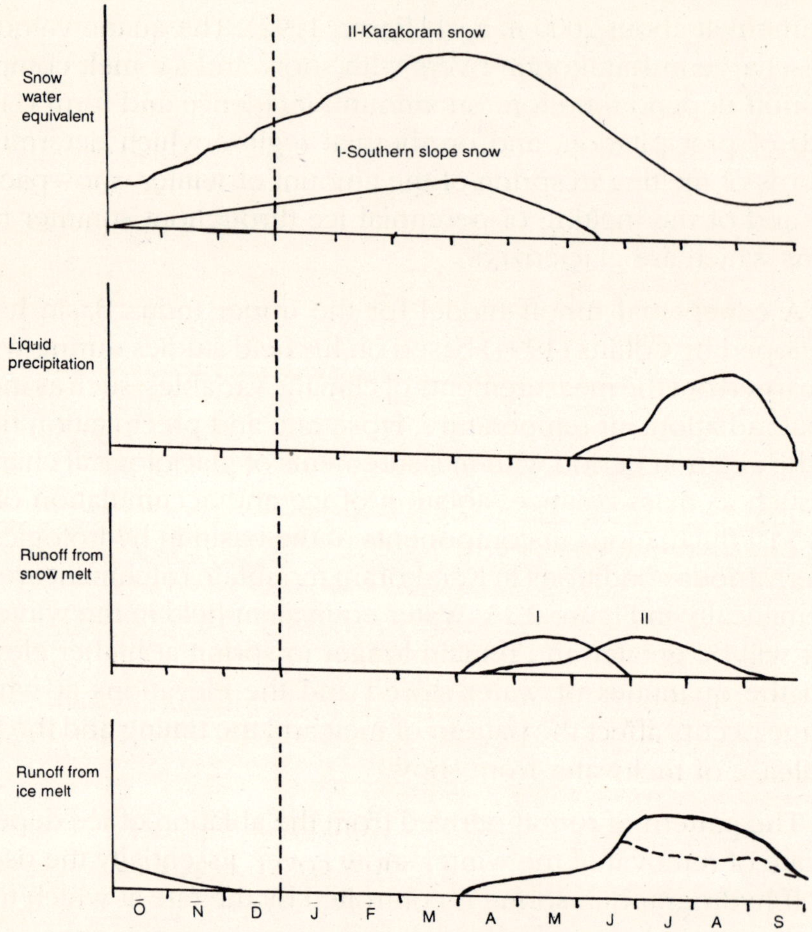

Fig. 2.13. Seasonal fluctuations in runoff components of the basin in hydrological cycle under various conditions in Karakoram mountain catchment. Reprinted from Collins (1994)

with high-albedo snow so that much of the increasing radiation through May and early June was not utilized in melting. Once snow cover is removed, the lower albedo of underlying ice leads to greatly increased specific yield of meltwater for the same energy input. Falling radiation inputs from August onwards, result in a generally downward trend in runoff.

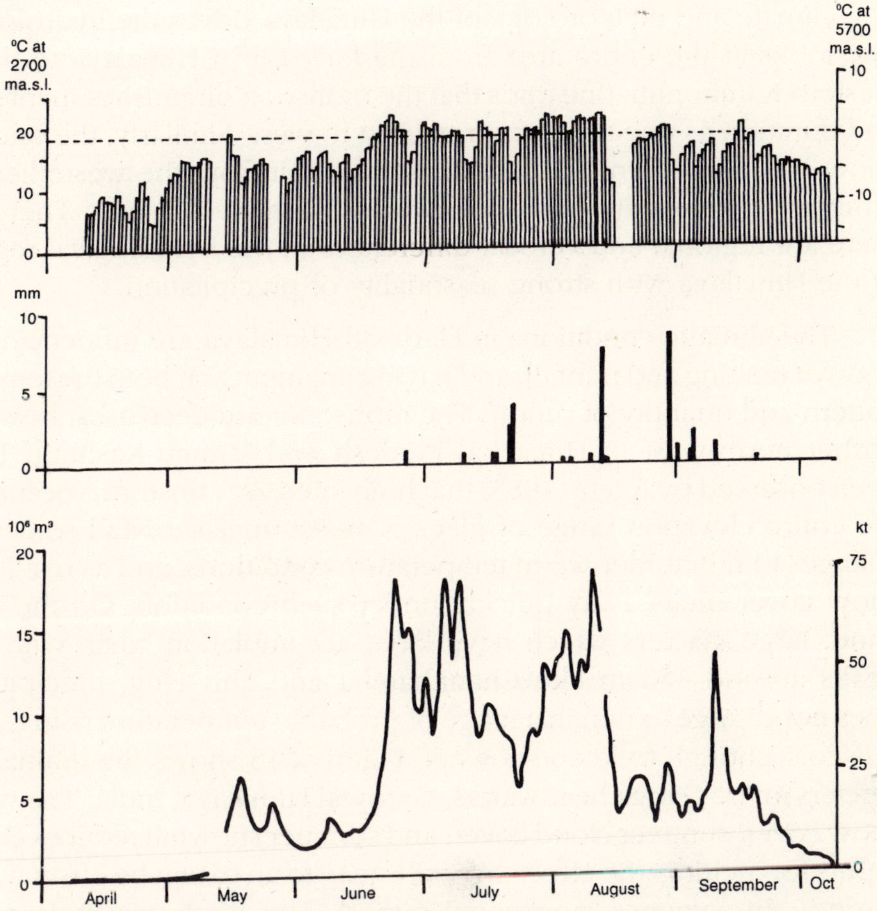

Fig. 2.14. Discharge, daily total precipitation and daily mean air temperature variations of the Batura glacier, Karakoram mountains, Pakistan. Reprinted from Collins and Hasnain (1995).

The runoff regime of the Hunza river at Danyore bridge reflects the pattern of discharge from the large glaciers in the basin, peaking in July or August. Ten percent of the flow in the Hunza is derived from the 4.9 % of the basin forming the Batura sub-basin (Table 2.1). Given minimal precipitation at lower elevations, that ten percent is probably derived from the 2.7 % of the Hunza basin occupied by Batura glacier itself. This indicates that glaciers are responsible for all the runoff (Collins, 1996).

2.11 HYDROLOGIC MODEL FOR GARHWAL HIMALAYA, INDIA

The climate and meteorology of the Himalaya drives the hydrology. If one looks at the entire area from the far eastern Himalaya to the far western Karakoram, one finds that the monsoon diminishes in intensity and duration as one traverses from east to west. Similarly, the seasonal snow and permanent ice cover mainly result from the westerlies and diminish in importance as one traverses from west to east. Therefore, there are regional and vertical differences in the hydrological regimes of the Himalaya with strong seasonality of precipitation.

The climatic conditions in Garhwal Himalaya are influencing energy for melting and rainfall, and having an impact on both the temporal pattern and quantity of runoff. The monsoon rain decreases, however, further north-west in Himachal Pradesh and Jammu-Kashmir. It has been observed by Ageta (1983) that high intensity rainstorms occur over the entire elevation range of glaciers as summer snowfall sometimes changes to rain under warm temperature conditions, and as result new snow cover melts away quickly under such condition. On the other hand, large glaciers which have large accumulation basin with high peaks around 6500m like Chaukhamba and Shiv-Ling, precipitation does not change to rain in a range of probable temperature rising, since it is cold enough to keep snowfall. Figure 2.15 shows the drainage of glaciers in the Ganga headwaters, Garhwal Himalaya, India. The winter snow cover, summer cloud cover, and summer snowfall reduces energy inputs for melting, the effect on runoff may to an extent be offset by rain during the summer monsoonal period. The discharge hydrograph, rainfall record, and air temperature record on Dokriani glacier, Garhwal Himalaya during the ablation season of 1994 is shown in Figure 2.16. The pattern of runoff first follow the snowmelt as the transient snowline receded from the non-glacier area and later on the glacier covered area where the recession process is uniform and slow. As the summer advances the transient snowline receded well up to the equilibrium line and this result in discharge peak by the end of June and early July. However, from mid-July to early September thick monsoonal cloud cover over the entire Garhwal Himalaya cuts off the radiation and the melting of ice substantially reduced and the discharges were mainly

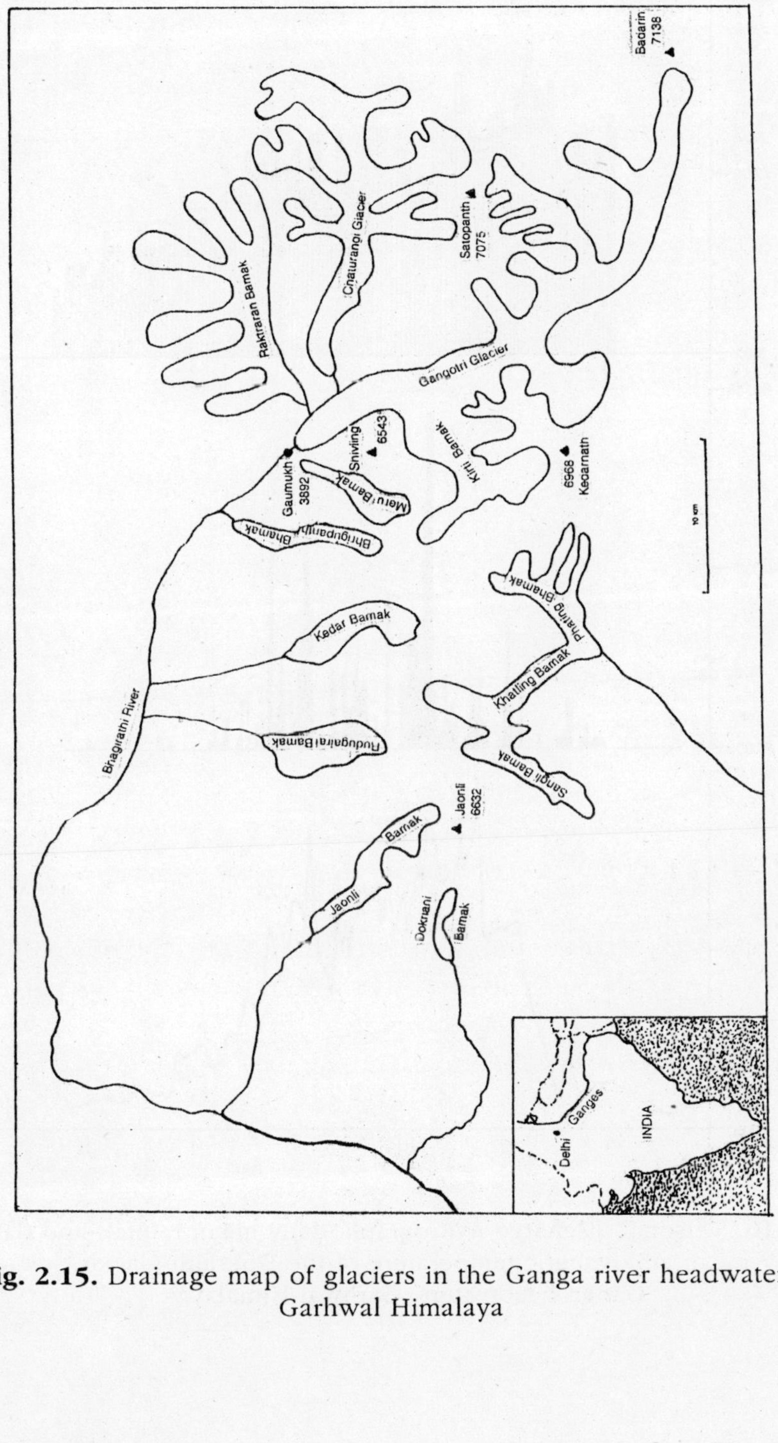

Fig. 2.15. Drainage map of glaciers in the Ganga river headwaters, Garhwal Himalaya

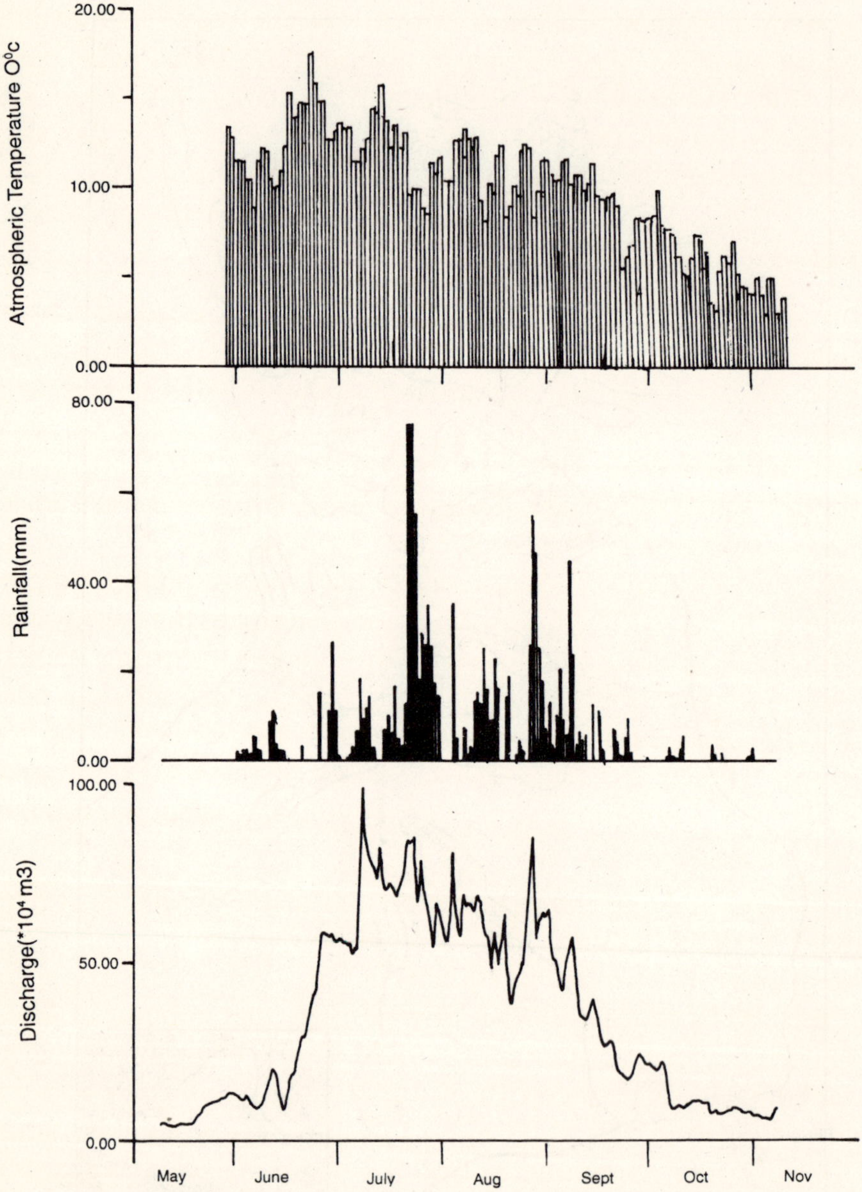

Fig.2.16. Seasonal discharge hydrograph, daily mean rainfall and daily mean atmospheric temperature of the Dokraini glacier, Ganga headwaters, Garhwal Himalaya

sustained by the runoff generated by the monsoonal rain. As a consequence of radiation interception by the cloud cover the resulting hydrograph shows a very distinct bite during the monsoonal period between July and September (Collins and Hasnain, 1994). From mid-September-November the discharge gradually reduces as the atmospheric temperature steadily falls.

2.11.1 Hydrological Processes

The Ganga river basin, having an area of 30,000 km^2, is called Garhwal Himalaya. The Garhwal Himalaya contains more than 1020 large and small glaciers. The basin has extreme variability in relief, precipitation and energy input. This is reflected in the diurnal and seasonal variations in climate and hence the variations in hydrology. Maximum flow takes place in June when the seasonal snowcover melt peaks as the transient snowline ascend upwards. The glacier melt is reduced, however, between July and September because of extensive monsoonal cloud cover over central Himalaya. The monsoonal period reduces the energy availability for melting during this period, the rainfall occurring over the entire elevation range nevertheless produces additional discharges which are not fully compensating the glacier melt component in the total hydrograph. Hence a distinct bite is visible in the total hydrograph obtained for the Langtang glacier, Nepal Himalaya as shown in Figure 2.17.

The Bhagirathi and the Alaknanda, two major proglacial streams, constitute the main drainage. Both streams join at Devaprayag to form the River Ganga. The proglacial streams of the Garhwal Himalaya carry about 70 to 80 % of their annual flow during the summer monsoon months (Bruijnzeel and Bremmer, 1989). The average rainfall in the Garhwal Himalaya is between 1000 to 2500 mm, of which of 80 % falls during the monsoon period between July and mid-September. Similar behaviour of runoff components from Dokraini glacier, has also been observed as shown in Figure 2.17. This indicates that the entire southern Himalaya inclusive its glacierized basins have a profound influence of monsoonal rainfall between July and September. Hence, the total hydrograph is composed of snowmelt, glacier ice melt, and monsoonal

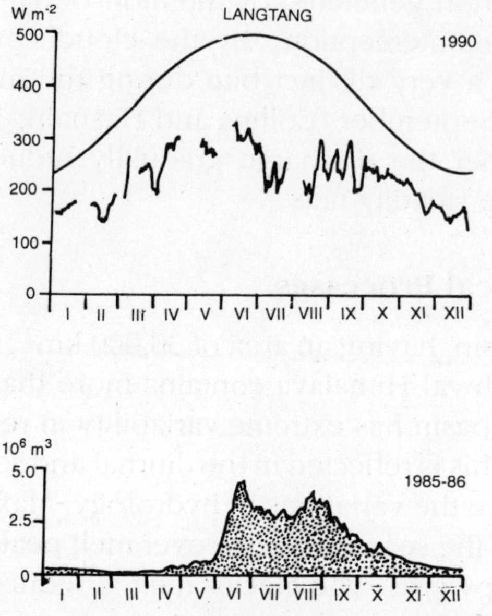

Fig. 2.17. Seasonal discharge hydrograph and average measured radiation at Langtang in the Nepal Himalaya. Reprinted from Fukushima *et. al.* (1987)

rainfall components. The seasonal and yearly variation of each component are linked to the variations in regional and local climate. When compared with drier west glacierized regions like Batura glacier, Karakoram Himalaya, Pakistan (Collins, 1995) the hydrograph does not show any bite during monsoonal period between July and September indicating the absence of monsoonal cloud cover. The total hydrograph characteristics of Batura glacier, however, resemble with those measured in the glacierized basins in Swiss Alps (Collins, 1989).

2.11.2 Monsoonal Control on Glacial Meltwater Discharge

Monsoonal precipitation over the glaciated basins has multiple role in controlling the hydrologic regime of the basins. Monsoonal cloud cover reduces the radiation input and thus lowering the glacier icemelt. The effects of subdued icemelt and monsoonal precipitation disturbs the rhythmic variations in the diurnal discharge. Such rhythmic variations in the diurnal discharge is considered as characteristic to the glacial discharges (Collins and Young, 1981). It is observed that the shape of the diurnal hydrograph is influenced by various parameters notably by the intensity and amount of precipitation at different elevation levels. Low intensity rainfall events, extended to long duration are very difficult to differentiate from the icemelt component. Whereas, storm events produces peak on diurnal hydrograph. Such multiple peaks on diurnal hydrographs often result from the variations in the radiative energy input and due to the monsoonal cloud cover. Water derived from the rainfall at the upper elevation regions of the glacier mixes with the supraglacial meltwater and routed through the glacier channels. Hencc, the characteristics of bulk flow hydrograph is more representative of glacier drainage characteristics than sources. The high intensity monsoonal rainfall over the lower elevation regions of the glacier directly influence the shape of the hydrograph, mainly due to rapid movement of water through the supraglacial and englacial channels.

2.11.3 Separation of Monsoonal Component

Runoff derived from the glaciated catchment, induced by the rainfall can be divided into two components: one is the direct contribution of liquid water from atmosphere to the system; and the second is enhanced ice and snowmelt due to the monsoonal precipitation. The icemelt induced by the rainfall events is considered to be minimal and can be accounted with total icemelt component. The extreme variability in precipitation over the Dokriani catchment is evident from the rainfall records of 1994 observation period. As expected the amount of rainfall over a region of high variability in relief and topography is entirely dependent on the characteristics of 'topoclimatic zones' existing within the catchment. Analysis of precipitation data from four meteorological observations has not shown any particular pattern of variation. Wind direction and velocity prevailing during the storm period may be playing a major role in determining the nature of distribution of precipitation over the basin. In the present study, data from these four stations representing varying topoclimatic zones of the basin have been used. Considering the small size of the catchment (10 km^2) four observatories are considered to be sufficient enough to represent the variations in the rainfall distribution over the glacierized basin.

The mass balance relationship has been used here to separate the daily rainfall component from the total discharge observed at the gauging site. The total area of the basin is divided as glaciarized and non-glaciarized area. It is necessary because the runoff characteristics are different for both regions. The liquid precipitation falling over the glacier is analogous to the channel precipitation, it joins the supraglacial meltwater channels and routed to the glacier portal through the glacial drainage system without major losses. But the rainfall over the non-glaciarized area of the basin account for the major losses mainly due to infiltration. Considering steep bare slopes of non-glaciarized part of the basin, a runoff coefficient of 0.7 (Mutreja, 1986) is considered to to be good for the calculations. The loss due to evaporation is considered to be negligible in such low temperature conditions. The areas of higher elevations experience solid precipitation during the monsoonal months. This depends on the atmospheric temperature prevailing at the time of storm occurrence. By using the temperature fall rate of 1.8°C/100 m, calculated by using the temperature data from four meteorological observatories, temperature at different altitudes has been calculated and found that area above 5400-5500 m experiences temperature $\leq$ 2°C for most of the days and hence is considered to be experiencing solid precipitation. Corresponding area has been deducted both from glaciarized and non-glaciarized regions for the calculations. Average basin precipitation has been calculated by taking the arithmetic mean of daily precipitation values from the four stations. The following relationship has been used to separate the rainfall component from the total daily glacial discharge:

For glaciarized region $R_{pg} = P \times A_g$

And for non-glaciarized region $R_{pn} = P \times A_n \times 0.7$

$$R_{pT} = R_{pg} + R_{pn}$$

where R_{pg} = Rainfall contribution from glacierized zone

R_{pn} = Rainfall contribution from non-glacierized zone

R_{pT} = Total daily rainfall contribution (m^3)

P = Precipitation in mm

A_g = Area of glacierized zone of the catchment where temperature > 2°C in m^2

A_n = Area of non-glacierized zone of the catchment where temperature > 2°C in m^2

Total catchment area of Dokriani glacier is 10 km^2 of which 6 km^2 covered with permanent snow and ice in which 4 km^2 experience liquid precipitation. Total non-glaciarized area is 4 km^2 and 2 km^2 experience liquid precipitation during the monsoonal period. Using this method the separated daily rainfall component is shown in Figure 2.18. The total rainfall experienced by the Dokriani basin during the 1994 ablation period between May and October was 1269 mm. With the advance of monsoonal period the cloud cover effectively reduces the solar energy input reaching the ground and hence reduces the melt rate. Recessional trend shown by both total hydrograph and corrected hydrograph for rainfall component for the monsoonal months proves that the reduction in melt rate due to the reduced energy input is not compensated by the rainfall as observed by Collins and Hasnain (1994) for Langtang glacier.

The relationship between atmospheric temperature and discharge has been studied. Correlation between daily average temperature and daily total discharge from May to October is poor ($r = 0.54$). It is improved to 0.62 with the rainfall corrected discharge. This relationship further improves between July and October. The total outflow from Dokriani basin during the six month (May-October) period of 1994 was 62.38×10^6 m^3 in which 11.46 % was contributed from monsoonal rainfall, which amounts to 7.15×10^6 m^3. Monthly contributions of both the components are given in the Table 2.2.

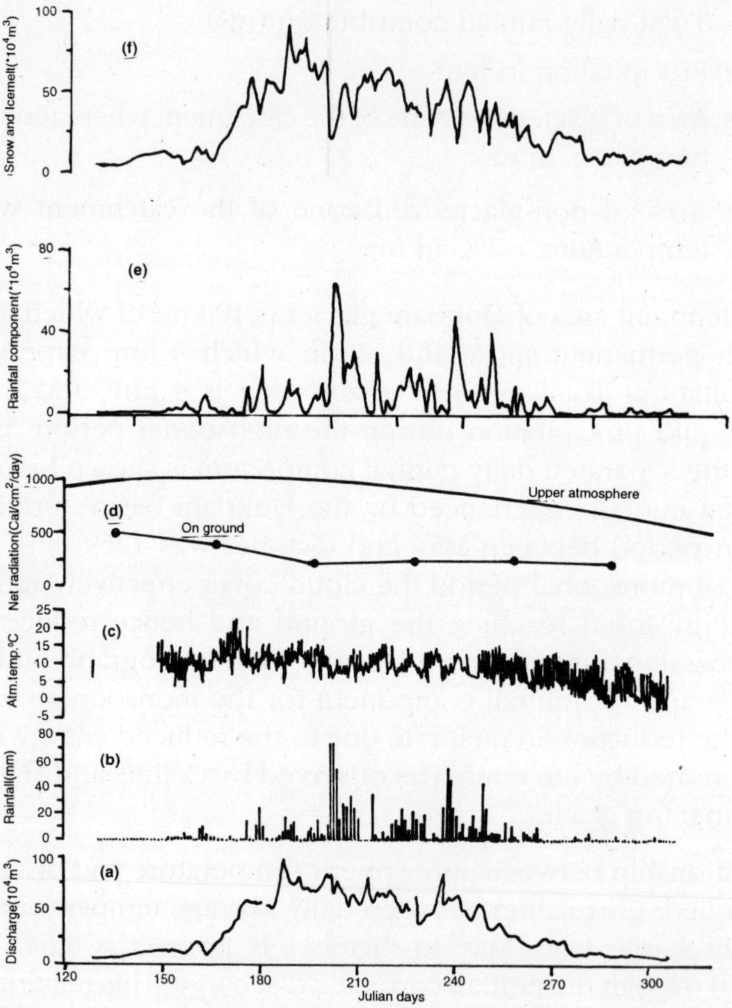

Fig. 2.18. (a) seasonal hydrograph; (b) seasonal mean daily rainfall; (c) seasonal mean daily atmospheric temperature; (d) average measured radiation on ground; (e) separated rainfall component; (f) separated snow and icemelt component of the Dokraini glacier; Ganga headwaters, Garhwal Himalaya.

Table 2.2. Monthly glacier runoff components during 1994 ablation period

Month	Total discharge 10^4 m^3	Rainfall component 10^4 m^3	Snow and Ice 10^4 m^3
May	195.52	2.141	193.38
June	867.89	64.9	802.99
July	2187.56	281.69	1905.87
August	1784.68	244.8	1539.88
September	909.96	106.59	803.37
October	292.46	14.87	277.59
Total	**6238.07**	**715.0**	**5523.08**

Hence, the runoff characteristics of glaciers in the Garhwal Himalaya are highly influenced by the monsoonal activity. On seasonal time scale it appears as a distinct bite on the crest segment of the ablation hydrograph, and suggests that the reduction in meltwater production induced by the monsoonal cloud cover is not fully compensated by the precipitation. The response of rainfall events on bulk flow hydrograph is controlled by the intensity and distribution pattern of monsoonal rainfall as well as the glacial drainage characteristics. When contributions from rainfall reach at the glacier portal, storm characteristics were modified by the glacier drainage characteristics. However, response of high intensity rainfall events at lower part of the ablation region retains the storm characteristics, while passing through the subglacial channel system. The separation of rainfall component from the total hydrograph is found to be useful for better understanding on controls of monsoonal activity on discharge characteristics of glacierized catchment. The monsoonal precipitation, and glacier drainage variations are perhaps the major factors controlling the runoff characteristics of glacierized catchments in the Ganga headwaters, Garhwal Himalaya.

3

GLACIAL HYDROCHEMISTRY

3.1 INTRODUCTION

Glacial hydrochemistry deals with the chemistry, and the chemical processes affecting glacier meltwater. This meltwater may be found above, within, beneath or in front of the glacier. It examines how meltwater evolves chemically as it is routed through a glacial system. An understanding of glacial hydrochemistry depends upon a thorough understanding of glacial hydrology. Thus, the glacial hydrochemistry and glacial hydrology are thought to play key roles in limiting the movement of glacier. Subglacial chemical processes may inhibit the velocity of basal sliding. Subglacial water storage in cavities may locally decrease local basal stresses, increasing basal sliding velocities. Draining of such cavities along subglacial hydrologic networks can slow glacier sliding, as the glaciers move by a combination of internal deformation and basal sliding.

The pioneering work of Hallet, 1976; Hallet *et. al.* 1978 has shown detailed thermodynamic calculations providing a considerable insight into subglacial chemical processes that have yet to be described by direct observations. Taking a different approach, Collins (1978; 1979; 1980) has pioneered the use of electrical conductivity measurements of glacial melts throughout the year, as a "surrogate measure" of the chemical load carried in proglacial streams. Collins has very successfully attempted to use the electrical conductivity of meltwater, along with discharge measurements, as an indirect way to trace the routing of different melt sources. Controls on meltwater composition are similar to those for other low-solute content waters. These waters have a low buffer capacity, which allow pH to vary widely in response to the local chemical environment of a water sample.

The hydrochemical research in glacial meltwaters has two main aims. The first has been to characterize water quality and quantity of geochemical weathering (Rainwater and Guy, 1961; Keller and Reesman, 1963; Reynolds and Johnson, 1972; Slatt, 1972; Church, 1974; Eyles *et al.*, 1982; Hasnain *et al.*, 1989). These studies show high rates of chemical weathering, as a result of rapid dissolution of freshly ground glacial rock flour. The second aim has been to identify chemical signatures which may separate runoff components and hydrological pathways through glacierized basins (Collins, 1978; 1979a; 1979b; 1981; Tranter and Raiswell, 1991; Brown and Tranter, 1990). The flow routing studies make the assumptions that (i) individual runoff components have constant compositions and (ii) they exhibit conservative mixing with respect to electrical conductivity.

3.2 PROCESS OF SOLUTE ACQUISITION

Solute is defined as all ions including base cations (Ca^{2+}, Mg^{2+}, K^+, Na^+), anions ($HCO_3^-, SO_4^{-3}, NO_3^-, Cl^-$), aqueous H^+, and neutral species (O_2, N_2, CO_2, H_4, SIO_4) dissolved in water. Solute acquisition (or chemical weathering) is therefore concerned with reactions which affect species in all phases: Solid, liquid and gas.

Therefore, solute acquisition is controlled by equilibria which depend on the availability of hydrogen ions, supplied from dissolution of atmospheric carbon dioxide and oxidation of sulphides (Raiswell, 1984). Carbon dioxide dissolution can supply H^+ at a rate required for consumption by weathering. Subglacial water is more reactive due to the presence of abraded and crushed sediment particles and this leads to rapid dissolution, when meltwater makes contact with basal morainic sediments or with the suspended sediment in transit. However, the rate of dissolution may not be high when large quantities of suspended sediment flowing through the subglacial conduits during summer time. Unreactive sediments would also lead to the low levels of solute content at times when sediment concentrations are high and rates of through flow of meltwaters from the glacier surface are high (Collins, 1979c). The potential for meltwaters to acquire solute in 'englacial' component is less because the short residence times for water on the ablation area surface before entering the ice-walled drainage system may reduce opportunity for equilibration of meltwaters with atmospheric gases. On the contrary, in 'subglacial' zone solute acquisition is greatest because there is slow transit of waters through an environment which promotes relatively rapid rates of chemical weathering. It seems that meltwater retarded subglacially is chemically-enriched to a greater extent than water flowing through the glacier more rapidly.

Dye tracing experiments carried out at the Haut Glacier d' Arolla, Switzerland by Brown *et. al.* (1994), indicate that the subglacial drainage system contains two principal components, a distributed system in which flow velocities are of the order of 0.05 ms^{-1}, and a channelized system in which flow velocities are in the range 0.5-1.0 ms^{-1}. During the melt season the channelized system expands headwards at the expense of the distributed system, so that by the end of the season it occupies the lowermost 3-3.5 km of the 4 km long glacier. Headwards growth of the subglacial drainage channel closely follows the upglacier retreat of the transient summer snowline on the glacier surface. This drainage configuration suggests that a two component model should adequately describe the subglacial transport of meltwaters.

Subglacial water composition is probably largely controlled by dissolution of carbonates and gypsum and by the weathering of alumino-silicates, as in natural waters in general (e.g. Stumm and Morgan, 1970; Morel and Hering, 1993). According to Raiswell (1984), oxidation of pyrite could release protons, which might in turn react with other minerals. Each of these processes will have an individual rate of reaction. Freshly-abraded, finely-divided products of glacial erosion will react readily with meltwaters. Small quantities of meltwater will acquire solute by percolating slowly through stored subglacial sediment. Considerable, if temporally-variable, concentrations of comminuted reactants are derived from channel margins along the entire length of basal hydrological pathways. These become suspended in meltwaters flowing through the network of interlinked subglacial cavities and conduits.

Collins (1995) has suggested that in the mid-to-late ablation season, much of the meltwater originating in the ablation zone flows a short distance across the ice surface in a supraglacial stream (Figure 3.1) before quickly descending a moulin (BC) to the glacier sole at C and

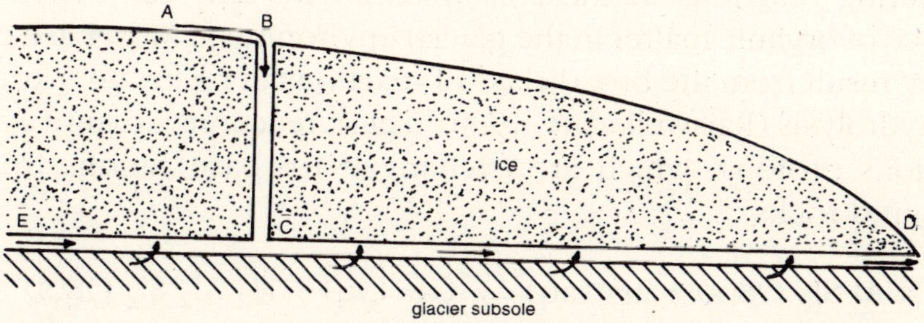

Fig. 3.1. Principal hydrological pathway of an Alpine glacier. Reprinted from Collins (1995).

reaching the portal (D) through a basal passageway. The glacier hydrological system consists of many such pathways ABCD, the basic unit of the plumbing. Between A and B gases are exchanged between atmosphere and meltwater. Sediment is entrained by parcels of water in transit from C to the portal and reactions between minerals and meltwater commence at C. Also, at C or between C and D, water from B may mix with meltwaters flowing from upglacier moulins (E) which already contain sediment and which will exhibit solute levels commensurate with the length of time in contact with minerals upstream. Some of this water will be from the snow-covered area and will have been delayed in the snow before reaching B. For snow melt, the distance CD will also be relatively large. For an entire glacier, discharge and solute content of meltwaters at the portal, at a particular point in time, are the sums of the volume and solute concentrations of the aliquots of water arriving at that time from all possible pathways ABCD, which will be of varying lengths, not only will there be a large of transit times for the various distances CD, making up the basal drainage net but that range will also vary with the overall level of discharge through the system. Both the absolute levels and daily ranges of measured portal solute concentration depend upon the reaction rate and the lengths of transit times, and their variability with diurnal discharge fluctuations.

3.3 SOURCES OF CATIONS

The water-rock interactions in the glacial system are quite similar to the weathering reactions in the soil system. The only difference is the absence of organic matter in the glacial environment. Most base cations usually result from the broad class of reactions in meltwaters knows as acid hydrolysis (Raiswell *et al.*, 1980). These reactions require a source of anions protons, which are exchanged for base cations from the powdered rocks.

$$Ca\ Al\ SI_2\ O8(s) + 2H^+(aq) \leftrightarrows Ca^{2+}(aq) + H_2\ Al_2\ Si_2\ O8(s)$$

Ca-feldspar Weathered feldspar surface.

$$CaCO_3(s) + H^+(aq) \leftrightarrows Ca_2^{2+}(aq) + HCO_3^-(eq)$$

Calcite

The aqueous protons required for acid hydrolysis are derived from two sources. The first is the dissociation of dissolved CO_2, derived via diffusion, either directly from the atmosphere or indirectly from gas bubbles within melting ice:

$$CO_2(aq) + H_2O(aq) = H_2CO_3^-(aq)$$

First dissociation

$$H_2CO_3(aq) = H^+(aq) + HCO_3(aq)$$

Second dissociation

$$HCO_3(aq) = H^+(aq) + CO_3^2(aq)$$

The solubility of carbon dioxide gas in water, therefore, contributes the neutral species carbonic acid and two anions—bicarbonate and carbonate. The relative proportion of these dissolved carbonate species depends on the values of the equilibrium constants K_{CO2}, K_1, K_2 and pH (- log aH) at a given temperature (Raymahashay, 1996). The dissolved carbonate equilibria at temperature 0°C the log K_{CO2}–1.12 and log K_1–6.5 and log K_2–10.62 (Garrels and Christ, 1965)

The second source of protons is the oxidation of sulphides, such as pyrite (Fe S_2), which are commonly found in the bedrock of mountain valley glaciers.

$$4FeS_2(s) + 15O_2(aq) + 14H_2O(aq) \leftrightharpoons 16H+(aq) + 4Fe(OH_3)2\,(s) + 8\,SO_4^{2-}\,(aq)$$

Pyrite

Oxidation of sulphides normally requires the presence of O_2, but in subglacial channels, where anaerobic conditions may develop, other oxidizing agents, such as NO_3^- may be used. Protons may also be derived from the dissolution of acids, either scavenged from the atmosphere by snow and rain (Barrie, 1991; Cadle, 1991). This may be the case in early melt season when snowmelt is usually more acidic (Tranter, 1991).

$$H_2SO_4(s) \leftrightharpoons 2H^+(aq) + SO_4^{-2}\,(aq)$$

acid sulphate aerosol

$$HNO_3(s) \leftrightharpoons H^+(aq) + NO_3^-\,(aq)$$

acid nitrate aerosol

Raiswell (1984) has stated that the major cation in glacier meltwaters is Ca^{2+}, which reflects the rapid dissolution of Ca^{2+} from silicate and aluminosilicate minerals.

3.4 SOURCES OF ANIONS

The major acid anions, Cl^-, HCO_3^-, NO_3^- and SO_4^{-2} are derived from various sources. It is assumed that Cl^- and NO_3^- are derived mainly from the dissolution of sea-salt aerosol (Souchez and Lemmens, 1987). The SO_4^{2-} is mainly derived from the oxidation of sulphides in powdered rock (Tranter and Raiswell, 1991). The two major anions in Alpine glacial meltwaters are HCO_3^- and SO_4^{2-} (Souchez and Lemmens, 1987), because the dissociation of atmospheric CO_2 and the oxidation of sulphides provide the bulk of the protons used to weather chemically carbonate, silicates and aluminosilicate minerals in glacial flour. Hence, the supply of atmospheric CO_2 and O_2 to meltwaters is a fundamental control on the rate and extent of chemical weathering in glacial meltwaters (Tranter *et. al.* 1993)

3.5. DOMINANT REACTIONS RESPONSIBLE FOR SOLUTE ACQUISITION IN MELTWATERS

3.5.1 Ion Exchange

Ion exchange is the rapid, reversible exchange of ions between solution and solid surface. Cation exchange is also a reversible exchange of positively charged ions. Ion and cation exchange by definition cannot add solute to meltwaters (Lemmens and Roger, 1978; Drewry, 1986). These reactions normally modify the relative proportion of ions present in dissolved state. The first reaction between water and powdered aluminosilicate and silicates usually involves a surface exchange of base cations for hydrogen ions (or protons) from solution (Wollast, 1967; Luce *et. al.*, 1972; Busenberg and Clemency, 1976; Lerman, 1979). The surface of the mineral may be physically altered during surface exchange rendering it difficult to replace or exchange the adsorbed protons for dissolved base cations (Garrels and Howard, 1956). Moreover surface exchange has a definite effect on the ability of the solution to dissolve carbon-dioxide. The loss of hydrogen ions from solutions increases the pH and as a consequence increases the quantity of dissolved carbon dioxide, bicarbonate, and calcite that the solution can hold for a fixed partial pressure of carbon-dioxide (Figure 3.2a). When isolated from the gas phase, the amount of dissolved CO_2 in the solution falls following surface exchange, since it is converted into HCO_3^- as shown in Figure 3.2b. Carbon dioxide diffusion from gas phase into the solution, result HCO_3^- and H^+ to dissociate from solution (Figure 3.2c). The hydrogen ions removed from solution, by surface exchange and the slower dissolution of the mineral lattice, are largely replaced by the base cations from rock-flour surface.

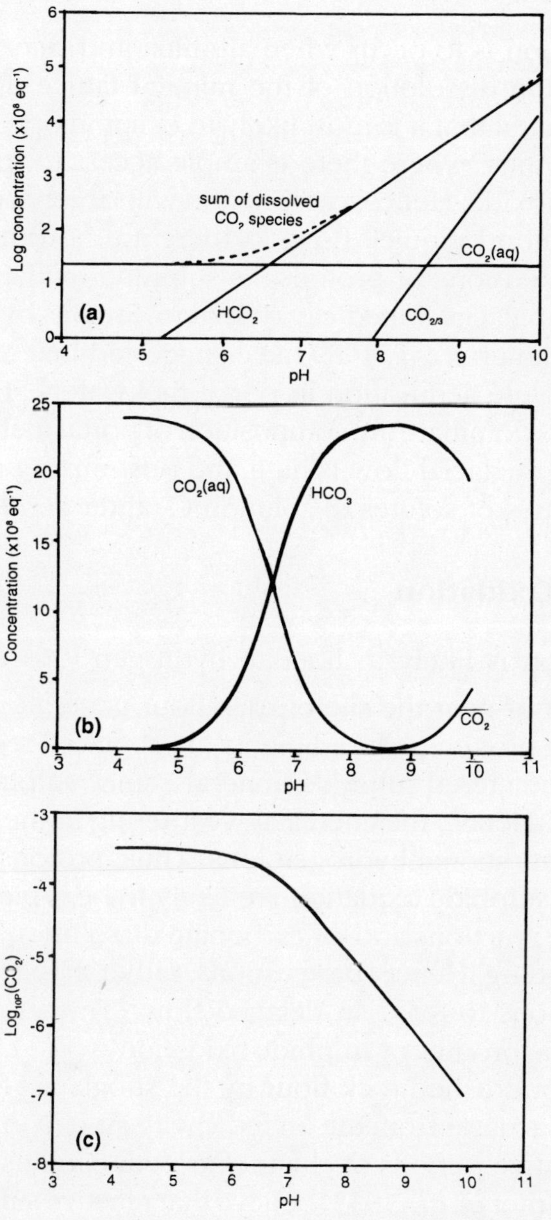

Fig. 3.2. (a) The relationship between the sum of dissolved carbon dioxide species $(CO_2)(aq) + (HCO_3^- + CO_3^{2-})$ and pH; (b) The effect of pH on the speciation of dissolved CO_2; (c) As the pH increases, CO_2 is converted into HCO_3^- or CO_3^{2-}. Reprinted from Tranter *et. al.*, (1993)

3.5.2 Carbonation

This type of reaction is to occur when alumina and silica are released into solution by the dissolution of the mineral lattice. Reynolds and Johnson (1972) stated that it is most likely to occur in open channels or partially filled cavities, where there is ample access to rock flour and gaseous carbon dioxide. Hence, carbonation will largely be responsible for solute acquisition by quick flow, as there are no other significant sources of hydrogen ions or protons, as sulphide oxidation is largely confined to the subglacial linked-cavity system. Studies by Raiswell and Thomas, 1984; Tranter *et. al.*, 1989 have suggested that carbonation is responsible for solute acquisition in portal meltwaters. Therefore carbonation reactions dominate the composition of portal meltwaters when the proportion of englacial flow is high and post-mixing reactions add significant quantities of solutes to solution (Tranter *et. al.*, 1993).

3.5.3 Sulphide Oxidation

Sulphide oxidation is likely to liberate hydrogen ions and SO_4^{2-}(i.e. Sulphuric acid) at or near the site of rock-flour generation (Tranter *et. al.*, 1993). Freshly broken rock sediments in subglacial region are most likely to contain unaltered sulphide minerals, since sulphide oxidation is relatively rapid. Hence, the chemical weathering in the linked-cavity system provides additional hydrogen ions. Thus, protons or hydrogen ions liberated by sulphide oxidation are likely to be rapidly utilized by surface exchange reactions and/or carbonate dissolution at or near the place of rock grinding. Hence, base cations, rather than hydrogen ions, are the counter ions to SO_4^{2-} in delayed flow (Tranter and Raiswell, 1991). Whereas the product of sulphide oxidation, SO_4^{2-}, can be leached from powdered or crushed rock flour by the subglacial flow, the rock flour sediment evacuates from the linked-cavity system in high concentrations during active periods of channel growth (Collins, 1989; Sharp, 1991). Hence, Tranter *et. al.* (1993) believe that reactive sulphides, either present as comminuted particles or close to the surface of the rock flour will be exhausted within the linked cavity system. The assumption that SO_4^{2-} is largely confined to subglacial flow (Tranter and Raiswell, 1991) is not very true during the early ablation season, when concentrated snowmelt may directly runoff the glacier surface and mix with portal meltwaters, consisting predominantly subglacial flow.

3.6 OPEN AND CLOSED SYSTEM SOLUTE ACQUISITION

The meltwaters thought to evolve chemically in open system when the atmospheric carbon dioxide $P(CO_2)$ in the gas phase is in equilibrium with the amount of dissolved carbon dioxide in the liquid phase; the system is to be closed when this is not the case. Open and closed system weathering regimes are kinetic phenomena and depend on the rate at which CO_2 diffuses into or out of solution relative to the rate of other chemical-weathering reactions (Raiswell, 1984; Raiswell and Thomas, 1984; Thomas and Raiswell, 1984). Figure 3.3 shows chemical evolution of meltwaters routing through open and closed systems.

3.6.1 Low $P(CO_2)$ System

In this system the rate of weathering is greater than the rate of hydrogen ion supply, the increase in pH induces CO_2 diffusion into solution, as the solution could hold additional dissolved carbon dioxide when large quantities of englacial flow mix with sediment laden delayed flow at the height of the ablation season (Sharp, 1991). Under these circumstances, the rate of proton consumption is greater than the rate of gaseous diffusion of CO_2 into solution.

3.6.2 High $P(CO_2)$ System

This may arise from three types of scenario. First an input of protons to the drainage system may occur, so as to give a short-lived reduction in pH, when acidic snowmelt mixes with a subglacial water body. This scenario is transient in most subglacial environment, since rock flour has a great capacity for absorbing protons. Secondly, the freezing of water body may result in an increase in $(CO_2)(aq)$, particularly in water filled cavities or channels when there is only limited opportunity for the solution to degas.

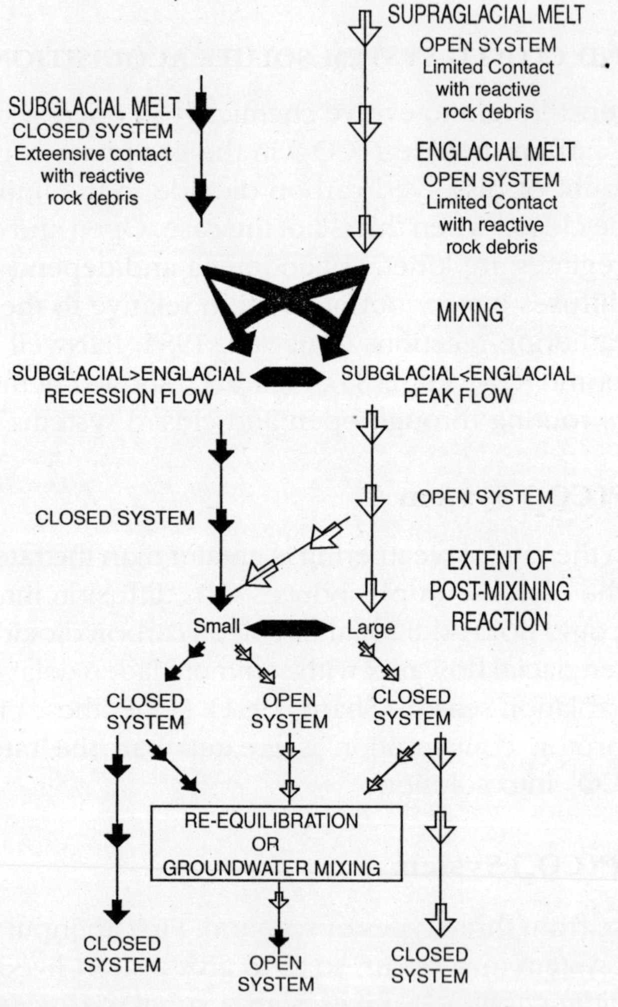

Fig. 3.3. Chemical evolution of meltwaters routing through open and closed system. Reprinted from Raiswell (1984)

Waters of the delayed flow contain both high concentrations of HCO_3^- and SO_4^{2-}. The mass fraction of sulphate is defined as the ratio of

SO_4^{-2} to the sum of SO_4^{2-} and HCO_3^-, where concentrations are in units of equivalents. This suggests that delayed flow, which dominates at lower discharge, contains a higher mass fraction of SO_4^{2-} than quick flow; further suggested that the atmospheric CO_2 of bulk meltwater increases as the mass fraction of SO_4^{2-} increases.

The combination of high concentrations of Ca^{2+}–HCO_3–SO_4^{2-} in the delayed flow in conjunction with elevated atmospheric CO_2 values tend to suggest that the processes of sulphide oxidation and carbonates dissolution are dominant controls on the composition of delayed flow.

$$\underset{\text{Pyrite}}{4FeS_2(s)} + \underset{\text{Calcite}}{16CaCO_3(s)} + 15\ O_2(aq) + 14H_2O(aq) \leftrightharpoons$$

$$16\ Ca^{2+}(aq) + 16\ HCO_3^-(aq) + 8SO_4^{2-}(aq) + 4Fe(OH)_3\ (s)$$

Studies on some of the Swiss glaciers also demonstrate that portal meltwaters at high discharge have a low mass fraction of SO_4^{2-} as shown in Figure 3.4. This indicates that these waters have low concentrations of atmospheric carbon dioxide. This may be due to post-mixing reactions.

Thus the solute acquisition processes in meltwaters stresses that dissolution and gas-diffusion kinetics are important controls on the composition of the two major discharge components and their port-mixing chemical evolution. These reaction kinetics are dependent on the residence time of waters within the channel and distributed system, the suspended sediment concentrations and the access of atmospheric gas supply—the factors constantly changing during the ablation season as the glacial drainage system evolves.

3.7 TWO COMPONENT MIXING MODEL

Total meltwaters composed of two components are termed the subglacial and englacial by Collins (1979b). Water flowing through the subglacial route, moves at slower rate, would become concentrate with ions on contact with the sediment-laden basal ice, bed rock, subglacial morainic sediments. The englacial water derived from the surface melt can flow

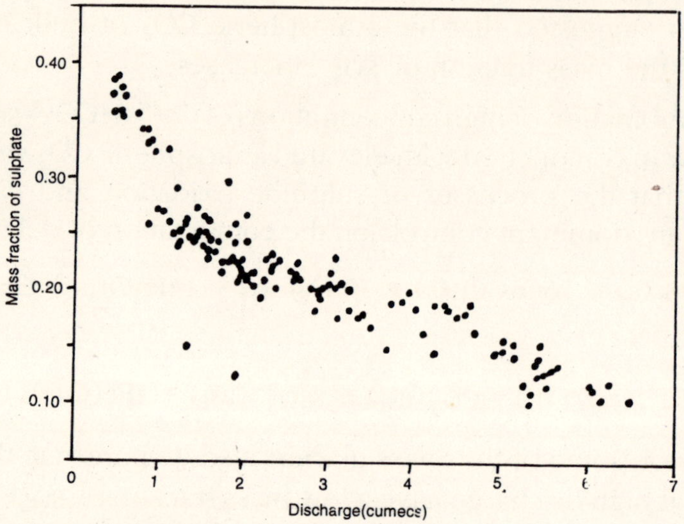

Fig. 3.4. The mass fraction of $SO_4^{2-}(SO_4^{2-} + HCO_3^-)$ as a function of discharge for bulk meltwaters draining Haut Glacier d' Arolla. Reprinted from Tranter *et. al.* (1993).

through the glacier in ice walled conduit system without delay and without chemical enrichment. A two-component mixing model describes the nature of water flowing in valley glaciers. The different solute loads in the two components are regarded as being essentially the result of the different residence times of the two components, the extent to which the water is in contact with reactive material, and its ability to sustain reactions, by the dissolution of oxygen and carbon dioxide into solution, together with the variability of hydrogen ions. The englacial component is regarded as that water which is transmitted relatively rapidly through mainly ice-walled conduits (Rothlisberger, 1972) whereas the subglacial component has a longer residence time, and is transmitted via a distributed linked-cavity system (Walder, 1986). Glaciers may have distributed drainage system or a combined distributed-channel system.

In the case of combined system, the distributed system drains into the channel system, particularly during the height of the ablation season, and there is a great potential for post mixing chemical evolution of portal meltwaters via interaction with elevated suspended-sediment concentrations. Before draining from the glacier portal, the two components mix in the major arterial channels at the glacier bed. It is only in these large subglacial channels that the bulk meltwaters encounter significant sediment concentrations. Brown *et. al.*, (1994) suggested that portal meltwaters may acquire a significant proportion of their solute load by so-called "post-mixing" chemical reactions, which involve the weathering of suspended sediments within this environment.

3.8 CHEMICAL WEATHERING PROCESSES

Glacier bedrock mostly contains silicates, aluminosilicates, and carbonates with accessory sulphide minerals. Sulphides break down rapidly in oxygenated surface environments and carbonate minerals generally weather more rapidly than silicates and aluminosilicates. Despite variations in the rate and directions of weathering, chemical evolution of melt waters always proceeds in the same direction and produces fixed quantitative relationship between the gain in cations and bicarbonate and the consumption of hydrogen ions.

The chemical weathering of rocks is one of the vital processes in the geochemical cycling of elements. The most important mechanism by which freshly ground rocks and minerals are weathered in glacial drainage system is by hydrolysis and acid hydrolysis reactions generally represent the most important subglacial chemical rock weathering processes (Raiswell, 1984, Tranter *et. al.*, 1993).

$$CaCO_3(s) + H^+(aq) = Ca^{2+}(aq) + HCO_3^-(aq)$$

The availability of free protons (H^+) to drive these reactions represents one of the primary controls over continued solute acquisition by glacial meltwaters. In addition to transient snowpack leaching (Tranter, 1991), protons may be derived from (I) the dissociation of H_2CO_3 (fueled by atmospheric CO_2); (ii) the oxidation of sulphide minerals such as pyrite represented respectively by:

$$CO_2(aq) + H_2O(aq) = H^+ + HCO_3^-(aq)$$

$$4FeS_2(s) + 15O_2(aq) + 14H_2O(aq) = 16H^+(aq) + 4Fe(OH)_3(s) + 8SO_4^{2-}(aq)$$

According to Hubbard and Nienow (1997), chemical reactions are partially coupled such that protons released by sulphide oxidation fuel accompanying carbonate hydrolysis:

$$4FeS_2(s) + 15\,O_2(aq) + 14\,H_2O(aq) + 16\,H^+(aq)$$
$$= 4\,Fe(OH)_3(s) + 8\,SO_4^{2-}(aq) + 16\,Ca^{2+}(aq) + 16\,HCO_3^-(aq)$$

It is clear from the above equations that the relative contributions of proton sources may be identified from the concentrations of bicarbonate and sulphate in glacial meltwaters. Brown *et. al.*,(1996) define the C-ratio as that of (HCO_3^- + SO_4^2) in equivalence units. A C-ratio of 1 therefore indicates acid hydrolysis that is fueled by protons provided by carbon dioxide dissolution (carbonation reaction) rather than by sulphide oxidation. In contrast, coupled weathering reactions result in a C-ratio of 0.5. Since, sulphide oxidation requires contact with reactive debris, and is therefore associated with distributed subglacial flows, while carbonation reactions require access to atmospheric carbon dioxide, and are therefore more strongly associated with open channel flow (Tranter *et al.*, 1993). Meltwater C-ratios analysis may provide information relating net flow pathways followed by total meltwaters. Brown *et al.* (1996) used this relationship to reconstruct temporal variations in solute acquisition at Haut Glacier d'Arolla, Switzerland. C-ratios were generally low (0.65) early in the melt-season and during recession flows, suggesting significant protons supply to the subglacial drainage system. In contrast, late melt-season flows were characterized by higher C-ratios (0.75-0.9), strongly suggesting a dominance of protons derived from atmospheric carbon dioxide during the periods when surface melting dominates glacier outflow.

Hubbard and Nienow (1997) suggested the relationship between the supply of protons to meltwaters and their consumption by weathering reactions may also be investigated on the basis of the partial pressure of carbon dioxide in solution in those meltwaters. $P(CO_2)$ is calculated as there is no direct field measuring device:

$$\text{Log } P(CO_2) = \log (HCO_3^-) - pH + PKCO_2 + PK_1$$

where concentrations are in mol l^{-1} and $P\,KCO_2 = 1.12$ and $PK_1 = 6.58$ (Ford and Williams, 1989). Raiswell (1984) developed this approach to characterize the chemical composition of total meltwaters in terms of flow through open or closed systems. In an open system the rate of proton supply to meltwaters matches that of their consumption by chemical weathering reactions. While the pH and alkalinity of such waters can increase with rock weathering, their $P(CO_2)$ remains in equilibrium with the atmospheric gas phase ($10^{-3.5}$ atm. at sea level). Conversely, a closed system is characterized by a mismatch between rates of proton consumption and supply. Tranter *et al.* (1993) stated that high-$P(CO_2)$ meltwaters could be generated in at least three situations: (i) where free protons are supplied to subglacial waters, via surface-derived acidic snow melt or the oxidation of sulphide minerals; (ii) where carbon dioxide is delivered to subglacial waters, and (iii) where meltwater acidity is neutralized by carbonate dissolution.

The subglacial weathering is therefore dependent on the flow pathways followed by glacial meltwaters. Inferences relating to those pathways may therefore potentially be made on the basis of diagnostic portal meltwater chemical signature.

3.8.1 Laboratory Dissolution Experiments

Laboratory dissolution experiments were conducted by Brown *et al.* (1994) to simulate chemical weathering by meltwaters under field conditions. They have chosen 3-h experiment time, based on due tracing studies; that is the transit time of a parcel of meltwater to pass through subglacial system. On a diurnal time-scale, it seems likely that there are two principal controls on the variability in composition of bulk glacial meltwaters. These are (1) the dilution of concentration flow by dilute quick-flow and (2) the extent of post-mixing reactions between bulk meltwaters and suspended sediment which will depend on the concentration of suspended sediment in the bulk meltwaters and on the duration of contact between these waters and suspended sediments. On a seasonal time-scale, additional factors include variations in the composition of the delayed flow resulting from changing residence times within the distributed component of the drainage system, and changes in drainage system structure which alter the relative proportions of quick-flow and delayed flow. Their laboratory experiments were designed to quantify the potential contribution of post-mixing reactions to this variability in bulk meltwater composition.

Measurement of pH and concentrations of Ca^{2+}, HCO_3^- and SO_4^{2-} from the dissolution experiments are presented in Figure 3.5. The amount of Ca^{2+} and HCO_3^- released, and protons consumed (denoted by increasing pH), increased continuously during each experiment and were positively associated with the concentration of sediment. However, the rates of increase in concentration and pH tended to decrease over time during each experiment. Similar patterns were found for the other base cations (Mg^{2+}, K^+ and Na^+). Final concentrations of Ca^{2+} and HCO_3^- (the equivalent ratio) of which approximated (1 : 1) ranged from 24 to 164 μequiv. l^{-1} after 3-h for experiments using suspended sediment concentrations of 0.3 and 4.0 gl^{-1}, respectively. Rate laws fitted to the experimental data for Ca^{2+}, HCO_3^-, and pH are presented in Table 3.1. Power curves were estimated by applying linear regression analysis to the $\log_e$, transformed dissolved ion concentration (dependent) and time (independent) variables and then back-transforming the estimated relationship (Williams, 1986). These laws resemble those presented elsewhere (e.g. Lerman, 1979). Similar rate laws apply to concentrations of Mg^{2+}, Na^+ and K^+, but these cations comprise 20% and usually 10% of the sum of base cations. Ca^{2+} and HCO_3^- are the dominant ions, and their approximately 1:1 ratio indicates that the carbonation reactions, examples of which are shown in the following chemical equations and characterize the dissolution experiments

$$\underset{\text{calcite}}{CaCO_3(s)} + CO_2(aq) + H_2O\ (aq) \rightleftharpoons Ca^{2+}(aq) + 2HCO_3^-(aq)$$

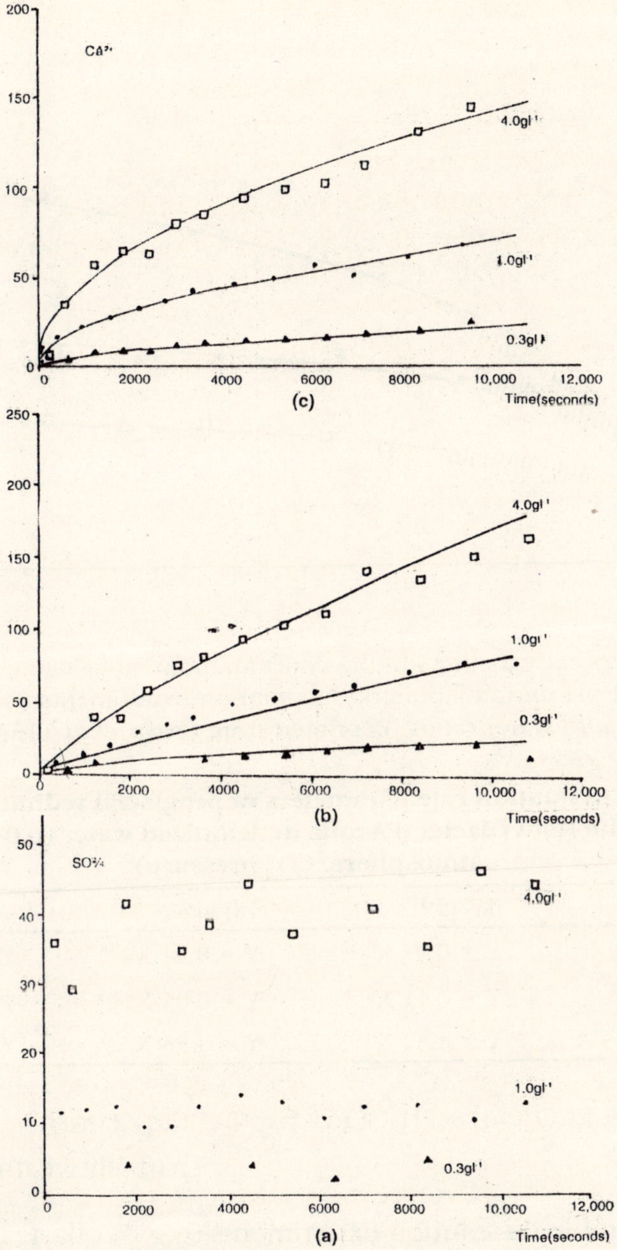

Fig. 3.5. Temporal variations in the concentrations of calcium, bicarbonate, sulphate and pH during laboratory dissolution experiments using different rock to water ratios. Reprinted from Sharp *et al.* (1995).

Fig. 3.5. *(Contd.)*

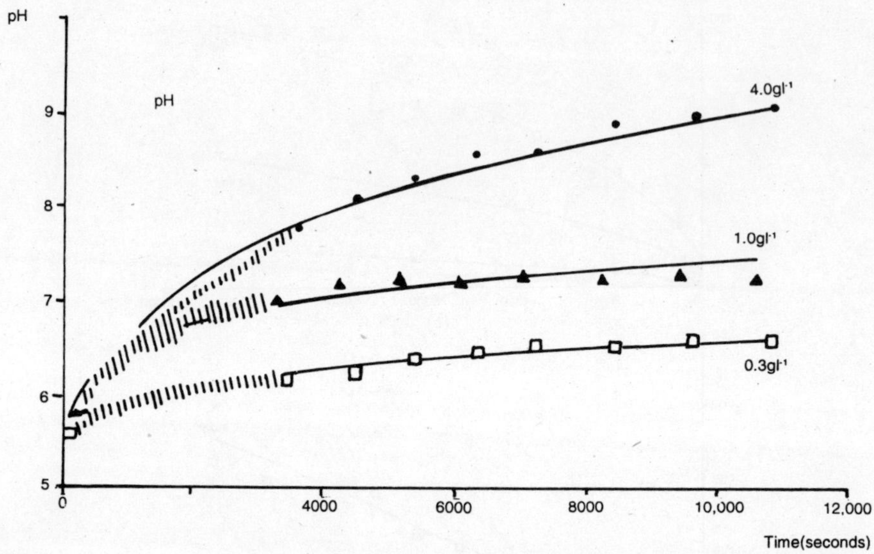

Fig. 3.5. Temporal variations in the concentrations of calcium, bicarbonate, sulphate and pH during laboratory dissolution experiments using different rock to water ratios. Reprinted from Sharp *et al.* (1995)

Table 3.1. Dissolution rate parameters of proglacial sediment derived from the Haut Glacier d'Arolla in deionized water (0.0-3.1°C, atmospheric CO_2 pressure)

	$0.3\ gl^{-1}$	$1.0\ gl^{-1}$	$4.0\ gl^{-1}$
Ca^{2+}	$Y = 0.07\ X^{0.62}$	$Y = 0.54\ X^{0.53}$	$Y = 1.55\ X^{0.49}$
HCO_3^-	$Y = 0.11\ X^{0.58}$	$Y = 0.05\ X^{0.80}$	$Y = 0.14\ X^{0.77}$
pH	$Y = 4.32\ X^{0.05}$	$Y = 4.53\ X^{0.05}$	$Y = 2.40\ X^{0.14}$

$$CaAl_2Si_2O_8(s) + 2CO_2(aq) + 2H_2O(aq) \rightleftharpoons Ca^{2+}(aq)\ 2\,HCO_3^- + H_2Al_2Si_2O_8(s)$$

anorthite — partially weathered feldspar

As in other simple dissolution experiments (e.g. Wollast, 1967), the pH increased immediately after the introduction of sediment to the system. The maximum pH values recorded ranged from 6.6 to 9.01 for experiments using 0.3 and 4.0 gl^{-1}, respectively. At 0.3 gl^{-1}, $P(CO_2)$ values deviated little from the atmospheric value; at 1.0 gl^{-1} they were lowered slightly below the atmospheric value, but at 4.0 gl-1 they were lowered significantly, reaching 10^{-5} atm after 3-h of contact. Lowering of the $P(CO_2)$ indicates that the consumption of protons in weathering reactions exceeded the supply by the diffusion of CO_2 across the gas-liquid interface. As the solution had free access to the atmosphere, this implies that the supply of protons to the dissolution system were kinetically controlled. This disequilibrium between the supply and consumption of protons was most marked at higher sediment concentrations, and tended to increase with time over the duration of the experiments. At all times the solutions were undersaturated in respect of calcite, though values of the saturation index (SI) tended to increase over time during each experiment and were higher in the experiments with higher sediment concentrations. Maximum values of $SI_{calcite}$ approached -0.7 at the end of the experiment using 4.0 gl^{-1} of rock flour.

There was some release of SO_4^{2-} during each experiment. Much of this release occurred immediately on addition of the suspended sediment to the solution. Thereafter there was minimal release. For example, in the 4 gl^{-1} experiment only 10 μequiv. 1^{-1} SO_4^{2-} was liberated after the initial release of 35 μequiv. 1^{-1}. This compares with the release of 150 μequiv.1^{-1} of HCO_3^- over the same period. They believe that the rapid initial release of SO_4^{2-} is an experimental artifact. During transport and storage, slow sulphide oxidation may occur at or near the mineral-grain surface. The SO_4^{2-} so formed is available for immediate leaching on contact with solution. Thereafter, the release of SO_4^{2-} is limited. This suggests that the release of sulphate by the oxidation of sulphides in suspended sediment is a slow process which would only be likely to contribute significantly to the chemistry of waters which spent long periods at the glacier bed. This supports the contention of Tranter *et al.* (1993) that sulphate is added to meltwaters primarily by pyrite oxidation which takes place within the distributed system, according to the following equation:

$$3FeS_2(s) + 15O_2(aq) + 14H_2O(aq) \rightleftharpoons 16H^+(aq) + 8SO_4^{2-}(aq) + 4Fe(OH)_3(s)$$

Thus, the experiments suggest that chemical weathering by dilute waters which spend relatively short periods in contact with suspended sediment is likely to be dominated by carbonation reactions. Ca^{2+} and HCO_3^- are the principal ions released by such reactions, their concentrations increasing as suspended sediment concentrations and contact time increase. Disequilibrium between rates of protons consumption and rates of proton supply results in lowering of $P(CO_2)$ to sub-atmospheric values, an affect which is most marked at higher sediment concentrations. Solutions remain undersaturated with respect to calcite at residence times and sediment concentrations typical of basal arterial channels. The release of sulphate by sulphide oxidation is minimal under these conditions.

David N. Collins, of the Alpine Glacier Project, University of Manchester, U.K. has performed another experiment to determine the rates of reaction of Findelengletscher, Zermatt Valley, Switzerland. The results of his studies are given by Collins (1995). He used icemelt-derived water which was allowed to equilibrate with the atmosphere at 2500 m as shown in Figure 3.6. The pH initially fell rapidly and then at a decreasing rate as dissolution of $CO_2(g)$ raised H^+ concentrating giving an equilibrium pH of 5.9 after about 8 min. In a further series of experiments, about 0.5 g of suspended sediment, collected wet after being deposited at the proglacial channel margin during high flow events and hence having had the shortest possible contact time with meltwaters, was added to meltwater which had been allowed to equilibrate with and left open to the atmosphere. Although rate of dissolution will depend on the concentration of sediment in the water, as indicated by Brown *et al.* (1994), a fixed concentration of about 2 kg m^{-3} was used in the experiments described here. The pH rose quickly to about 8.0 in the first 6-7 min. the rate showing pH 8.85 after 30 min. EC (a measure of total ionic strength and hence indicating the overall effect of dissolution and reaction between meltwaters and sediment) increased from the initial level of ice melt, 4 $\mu S\ cm^{-1}$, to 8 $\mu S\ cm^{-1}$ in 10 min., and, 10 $\mu S\ cm^{-1}$ after 25 min. with the growth rate continuing to decline, EC steadily increased to 17 $\mu S\ cm^{-1}$ after 180 min. and so on to about 30 $\mu S\ cm^{-1}$ after 5 h. In three experiments conducted by Collins over longer periods of time, EC

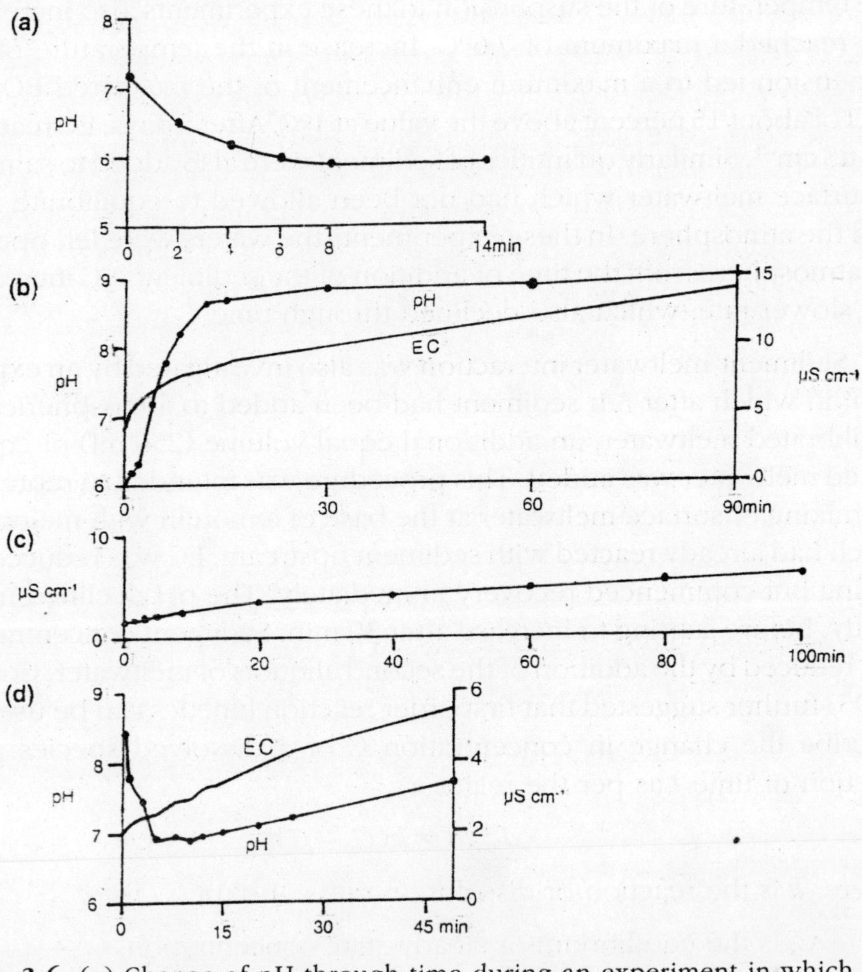

Fig. 3.6. (a) Change of pH through time during an experiment in which ice-derived meltwaters from the surface of the Findelen glacier equilibrated with atmosphere; (b) curves showing the increase of electrical conductivity and pH through time, after addition of suspended sediment to a sample of surface meltwater which had first been allowed to equilibrate with the atmosphere; (c) curve showing the increase of EC after addition of suspended sediment to sample of surface meltwater which had not been allowed to equilibrate with the atmosphere; (d) curves indicating the increase of EC and first the decrease and then increase of pH after addition suspended sediment to sample of surface meltwater which had not been allowed to equilibrate with the atmosphere and which was maintained closed to the atmosphere during the experiment. Reprinted from Collins (1995).

continued to rise, to 37 μS cm^{-1} after 10 h, and to 74.0 μS cm^{-1} after 24 h. The temperature of the suspension in these experiments also increased and reached a maximum of 7.6°C. Increase in the temperature of the suspension led to a maximum enhancement of the measured EC to a level of about 15 percent above the value at 1°C. After 5 days, EC reached 145 μS cm^{-1}. Similarly quantities of sediment were also added to samples of surface meltwater which had not been allowed to equilibrate fully with the atmosphere. In these experiments the waters were left open to the atmosphere from the time of addition of the sediment. EC increased at a slower rate, which also declined through time.

Sediment-meltwater interaction was also investigated by an experiment in which after 3 h sediment had been added to atmospherically-equilibrated meltwater, an additional equal volume (250 ml) of equilibrated meltwater was added. This procedure was intended to represent the mixing of surface meltwater at the base of a moulin with meltwater which had already reacted with sediment upstream. EC was reduced on mixing but commenced recovery immediately. The pH declined more gently, before starting to increase after 30 min. Sediment concentration was reduced by the addition of the second aliquots of meltwater. Collins (1995) further suggested that first order reaction kinetics can be used to describe the change in concentration C of a dissolved species as a function of time t as per the relation.

$$dc/dt = k(C_s - C)$$

Where: k is the reaction or dissolution rate constant

C_s is the equilibrium or steady-state concentration

dc/dt is a function of the distance of C from C_s (Lerman, 1979).

The dissolution model suggested by Collins (1995) appears to be more appropriate. After an initial rapid increase in EC on addition of sediment, the rate of weathering slowed down to give a steady continuing rise in EC. After about 45 min, rates of increase of EC of about 0.065 mS cm^{-1} for meltwaters fully-aerated and of about 0.035 for meltwaters partially-aerated at the start, both open to the atmosphere during the experiments, were obtained. Size of suspended sediment particles will additionally influence the rate of increase of EC with time.

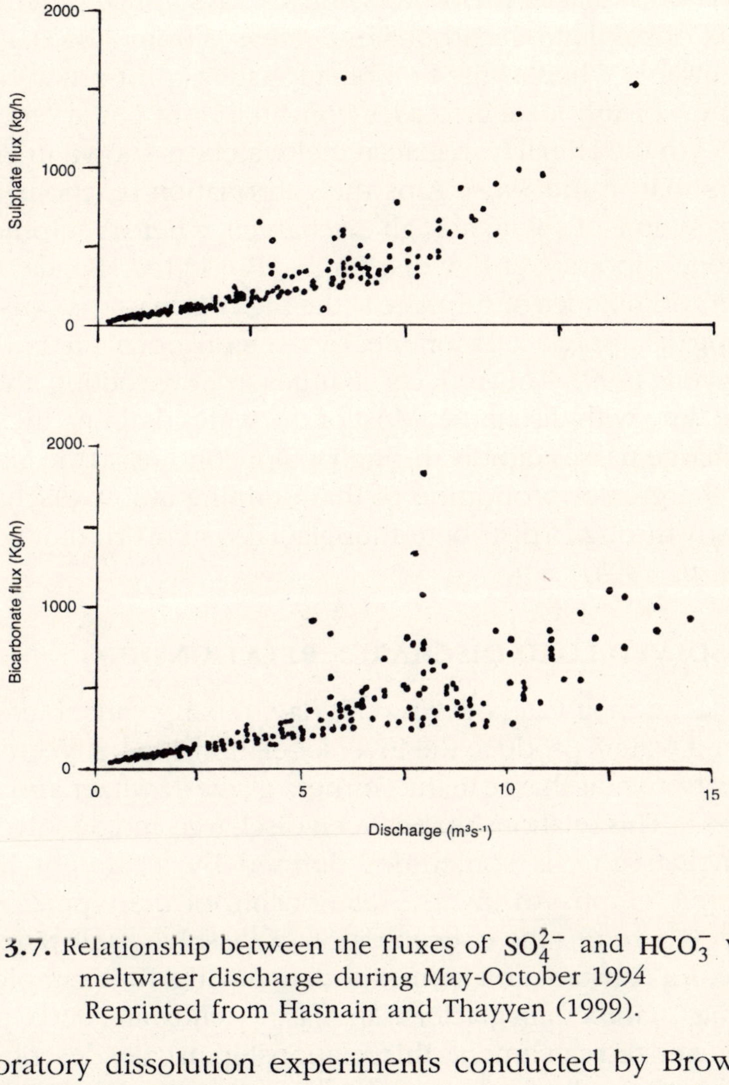

Fig. 3.7. Relationship between the fluxes of SO_4^{2-} and HCO_3^- with meltwater discharge during May-October 1994 Reprinted from Hasnain and Thayyen (1999).

Laboratory dissolution experiments conducted by Brown *et. al.* (1994) and Collins (1995) are not precise analogues of the type of processes occurring in nature.

3.9 CONTROLS ON SOLUTE FLUXES

Solute fluxes in glacial meltwaters usually increase with discharge. The largest solute fluxes are transported by the most dilute meltwaters as solute concentration is inversely related to discharge Sharp *et al.* (1995). In the Swiss Alps glacier meltwaters, the flux of sulphate as a proportion of the flux of sulphate+bicarbonate, decreases from 25% at 1 $m^3 s^{-1}$ to 12% at 6 $m^3 s^{-1}$. Whereas the flux of sulphate as proportion of the flux of sulphate + bicarbonate increases from 40-45% at 2 $m^3 s^{-1}$ to 50 to 63% at 10 $m^3 s^{-1}$ in the Himalayan glacial meltwaters as shown in Figure 3.7. It appears that in the Swiss Alps the carbonation reactions are more important source of solute at high discharges, whereas sulphide oxidation is more important at low discharges. But in the Himalayan glacial meltwaters dominance of sulphate at the high discharge is caused by the weathering of supraglacial moraines by the monsoonal rainfall (Hasnain and Thayyen, 1999). This reflects changes in flow routing and varying residence time with discharge. Most of the water draining the glacier at high discharge passes rapidly through major conduits at the glacier bed, whereas the greater proportion of that draining at low discharge pass more slowly through a distributed subglacial system (Tranter *et al.*, 1993; Brown *et al.*, 1994).

3.10 DISSOLVED LOAD-DISCHARGE RELATIONSHIP

Surface glacier meltwaters during the day, mixing with solute-rich basally routed waters produce the inverse relationship between electrical conductivity and discharge in the Gornera glacier, Switzerland as shown in Figure 3.8. This relationship between discharge and dissolved load in glacial meltstreams is sometimes defined by a straight line on a bilogarithmic graph. An inverse relationship of the type $C = kQ^{-n}$ is exhibited, where C is the concentration of dissolved substances, Q the discharge, k a constant and n an exponent having, for example, a value of 0.6 in the Chamberlain glacier as in Alaska (Rainwater and Guy, 1961). In non-glacierized catchments this relationship usually described by the model $c = aQ^{-b}$, where c is dissolved solids concentration (electrical conductivity) and Q discharge (Hem, 1970). This is the simplest of several mixing models (Hall, 1970). Simple models are not appropriate to describe glacier meltwaters runoff because of different sources and routings. Collins (1979) has proposed a model by the margins of trapezium as shown in Figure 3.9 based on his studies on Gornera glacier,

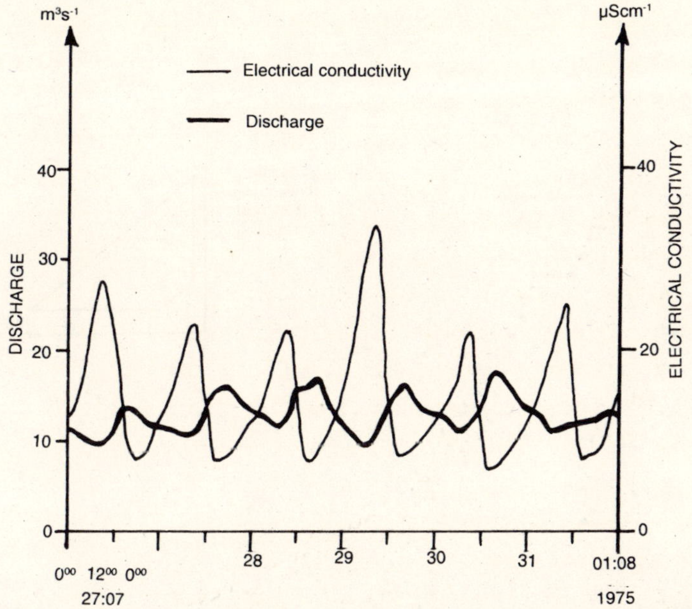

Fig. 3.8. Temporal variations of electrical conductivity and discharge in the Gornera glacier, Switzerland. Reprinted from Collins (1979a)

Switzerland. Side AB represents the lowest observed conductivity which occurs independent of discharge at higher flows, and which is determined by the solute content of ice and snow meltwaters from the glacier surface at times when there is minimal subglacial outflow. Point D is located by the maximum solute concentration which ideally is the concentration which would be measured in interstitial waters of basal moraine. Side CD is positioned by lowest observed recession flows which result from varying proportions of surface runoff and basal outflow. When further surface contributions inhibit basal outflow, solute content is decreased (CB), as discharge rises. Some water probably restocks subglacial storage within conduits following their depletion up to the discharge represented by point C permitting a wide range of water quality at an almost constant low flow. The upper limits of the distribution (AD) are located by daily maximum conductivities.

Points in the distribution close to AB represent meltwaters with chemical characteristics primarily derived from atmospheric inputs.

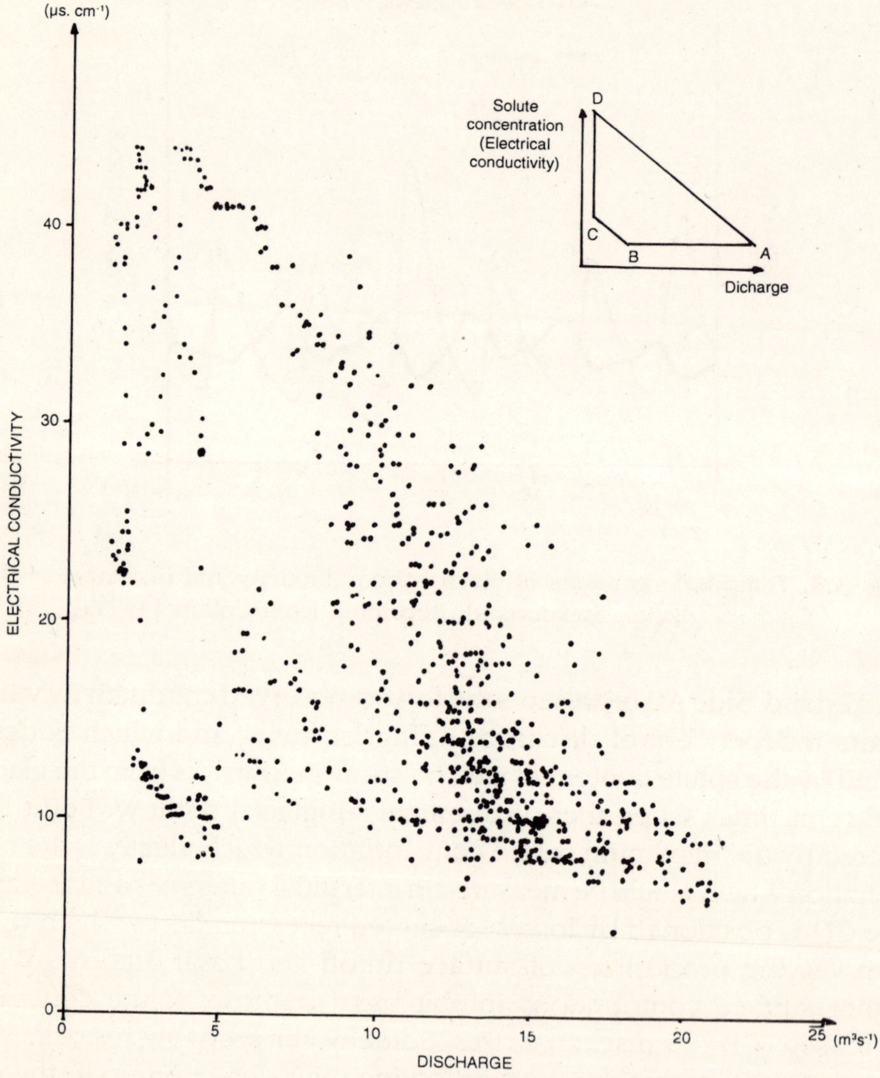

Fig. 3.9. Inverse relationship between solute concentration (as electrical conductivity) and discharge in the Gornera glacier. Inset: trapezoidal frame ABCD which encloses the distribution. Reprinted from Collins (1979a)

Towards D, terrestrial sedimentary solute origins become dominant. Different mixing models are appropriate for different parts of trapezium frame, according to pertaining hydrological conditions. In the case of the Tsidjiore Nouve glacier meltstream, equations have been obtained for the concentration of Na, K, Ca, Mg in samples collected in June, July and August. If discharge is expressed in $m^3 s^{-1}$ and concentration in μequiv. 1^{-1}, then:

for Na, $C = 11.22\ Q^{-0.59}$

for K, $C = 10.47\ Q^{-0.37}$

for Ca, $C = 263.02\ Q^{-0.54}$

for Mg, $C = 27.54\ Q^{-0.72}$

Thus, higher discharges were characterized by a lower concentration of the four cations. The correlation coefficients were significant at a confidence level of 0.999 and were, respectively, -0.83 for Na, –0.70 for K and 0.95 for Ca and Mg (Lemmens and Roger, 1978; Lemmens, 1978).

The concentration range is not connected with the suspended load. Indeed, variations of the concentration per liter of cations absorbed on suspended particles, obtained by multiplying the concentration per gram by the weight of suspended sediment per liter of streamwater, were an order of magnitude less than those for the corresponding dissolved cations expressed in the same units (Lemmens and Roger, 1978). Fluctuations of suspended load cannot, therefore, explain variations in dissolved cations.

Such a relationship can be explained by mixing of waters from two different sources. In a hydrological model proposed by Collins (1977, 1979a), mainly for the Gornergletscher, two main components of glacial discharge were recognized: a dilute, surface meltwater which flows rapidly with little alteration through englacial channels and solute-rich waters flowing more slowly through the subglacial system. A simple two component mixing model can be used to explain the temporal variations of the bulk meltwater emerging at the portal. The different slopes of the concentration-discharge regression lines for the four major cations can now be examined (Figure 3.10), bearing in mind that water at low discharge, essentially subglacial water, is closest to equilibrium with mineral particles. If the concentration of dissolved cations is controlled

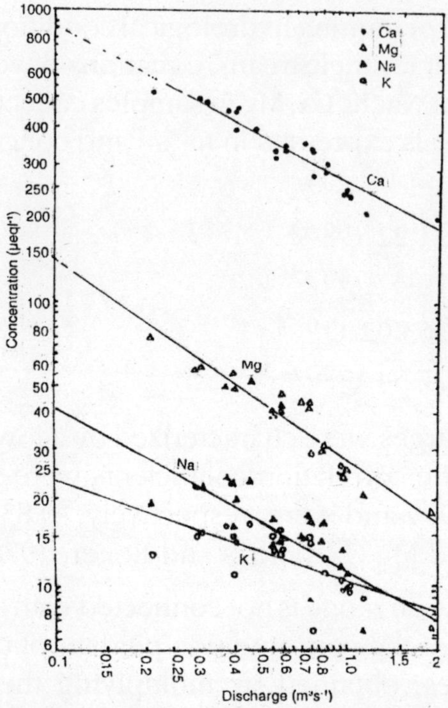

Fig. 3.10. Relationship between the concentration in the four major cations and discharge for the glacier de Tsidjiore Nouve meltstream. Reprinted from Souchez and Lemmens (1987)

by diffusion, the concentration ratio of a less mobile versus a more mobile cation (i.e. of a cation with a lower diffusion coefficient versus a cation with a higher diffusion coefficient) is lower when concentrations are distant from equilibrium values than when equilibrium values are approached. This concentration ratio must then increase when the discharge decreases. The slope of the regression line in the bilogarithmic coordinate system used in Figure 3.10 must, therefore, be steeper for a less mobile cation (i.e. a cation with a lower diffusion coefficient). Results of the experiments using fine particles from the Tsidjiore Nouve glacier basin indicate that K diffuses more quickly than Na and Mg is the slowest. The calculated values of the slopes of the regression lines are –0.72 for Mg, –0.59 for Na, –0.54 for Ca and –0.37 for K. The slope sequence for Na, K and Mg corresponds exactly to the sequence of the diffusion coefficients. The slope of Ca would be intermediate between those of Na and Mg if we consider the diffusion coefficient in water and this is almost the case with the Tsidjiore Nouve data. In fact, the slope for Na (–0.59) is slightly steeper than that for Ca (–0.54) but this difference is probably not significant.

The linear relationship, on a bilogarithmic graph, between discharge and dissolved load is not observed for all Alpine meltstreams studied. Indeed, Collins (1979b) studying the chemical composition of water emerging from the snout of the Gornergletscher, pointed out a great range of ionic concentration associated with a given discharge and showed clockwise hysteresis loops of conductivity and discharge on diurnal time scale.

The use of dye to determine transit times of meltwater from moulins to portal and the use of these transit times to characterize diurnal variation in the length of period of contact between suspended sediment and meltwater beneath the glacier allow a simple Lagrangian coupling with experimentally-determined rates of increase of EC through time. This model provides an indication of the level in diurnal variation of solute concentration that parcels of meltwater draining from the moulins in the lower ablation zone would exhibit on reaching the portal of Findelengletscher.

According to Sharp *et al.* (1995) glacierization of a catchment may increase the chemical weathering rate. The efficacy of chemical weathering in glacial environments is attributable to (1) high flushing rates, which ensure that dilute meltwaters do not reach chemical equilibrium with available sediment and the reaction rates remain high; (2) turbulent meltwaters, which transport large amounts of fine, freshly ground, chemically reactive rock flour; (3) high suspended sediment concentrations, which increases with meltwater discharge, thus increasing the flux of reaction sites; and (4) the low buffering capacity of dilute meltwaters. When such waters come into contact with large amounts of reactive material, available protons are consumed more rapidly then they can be replaced by diffusion of CO_2 into solution, causing the pH to rise, the $p(CO_2)$ to drop, and CO_2 to diffuse into solution.

The link between discharge and CO_2 drawdown suggests a potentially important role for glacially driven chemical weathering in carbon cycling over glacial-interglacial time scales.

3.11 FACTORS CONTROLLING SOLUTE ACQUISITION PROCESSES IN SOUTHERN HIMALAYAN MELTWATERS

Dokriani glacier, Ganga basin, Garhwal Himalaya, India is a smaller glacier, located in the Uttarkashi district of Uttar Pradesh has been studied with the objective to understand chemistry and chemical processes affecting glacial meltwater. The glacier lies 25 km east of Bhukki village. It extends from latitude 31° 49'N to 35° 20'N and longitude 78° 47' E to 78° 5' E (Figure 3.11). The total length of the glacier is 5 km. The stream emerging at the glacial portal is known as Din Gad, which joins with the Bhagirathi at Bhukki. The total area of the glacier catchment is 10 km^2, of which 6 km^2 is glacierized. Three quarter of the glacier ablation area is covered by thick debris. The accumulation area of the glacier is about 2.5 km^2, consisting of two cirque glaciers, one originating on Draupadi ka Danda and the other on Janoli. The equilibrium line is –4960 m.

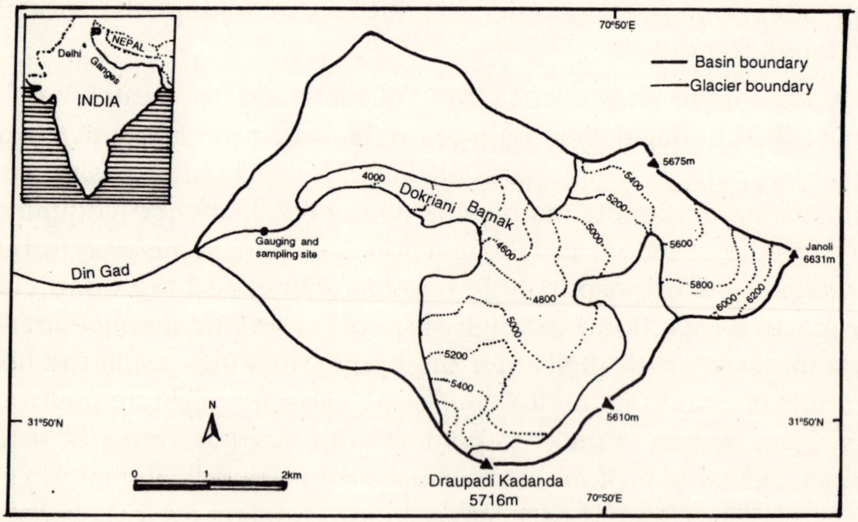

Fig. 3.11. Map of the Dokriani glacier catchment, Ganga headwater, Garhwal Himalaya. Reprinted from Hasnain (1996).

The chemical composition of Dokriani glacier meltwater for low flows (0800 Hrs) and high flows (1700 Hrs) for selected flow regimes during the 1994 ablation period is shown in Table 3.2 which also shows the percentage charge balance error. The charge balance between cations and anions is <10%, confirming the reliability of the analytical

Table 3.2 Major ion composition of the Dokriani glacier meltwater at selected discharges from 8 May to 31 October, 1994 (in meq l^{-1})

Date	Time	Julianday	Discharge m^3s^{-1}	Ca^{2+}	Mg^{2+}	Na^+	K^+	HCO_3^-	SO_4^{2-}	Cl^-	%diff.
8/5/94	0800	128.334	0.50	525	114	58	65	366	418	3	-1.56
	1700	128.708	0.60	458	92	49	60	286	279	7	7.02
24/5/94	0800	144.334	1.22	405	75	44	65	262	277	9	3.61
	1700	144.708	1.43	365	68	40	65	251	250	6	2.91
20/6/94	0800	171.334	3.88	307	49	22	48	246	191	3	-1.68
	1700	171.708	4.26	246	37	30	53	158	244	11	-6.14
29/6/94	0800	180.334	6.78	420	73	24	82	294	297	5	0.25
	1700	180.708	7.47	469	58	18	77	318	279	4	1.65
30/6/94	0800	181.334	5.26	1384	73	24	108	803	737	7	1.36
	1700	181.708	7.95	1941	72	20	128	1053	1140	3	-0.80
1/7/94	0800	182.334	5.68	990	65	24	96	673	557	3	-2.41
	1700	182.708	7.71	693	56	18	82	458	426	4	-2.25
2/7/94	0800	183.334	5.68	619	70	22	89	455	342	2	0.06
	1700	183.708	7.95	686	60	20	89	449	424	4	-1.29
3/7/94	0800	184.334	4.52	421	54	20	72	234	312	3	1.59
	1700	184.708	8.15	537	51	18	72	375	301	2	-0.05
24/7/94	0800	205.334	5.69	418	44	16	57	362	218	2	-4.16
	1700	205.708	12.50	603	42	13	72	403	349	2	-1.65
11/8/94	0800	223.334	5.02	283	45	24	45	205	171	2	2.37
	1700	223.708	7.93	271	34	18	47	185	190	3	-1.27
25/8/94	0800	237..334	4.57	319	62	26	59	198	266	2	-0.08
	1700	237.708	9.47	278	41	18	64	130	268	2	0.17
16/9/94	0800	259..334	2.44	373	68	34	52	275	197	2	5.29
	1700	259.708	3.92	253	44	21	41	184	139	1	4.78
8/10/94	0800	281.334	0.98	467	103	48	61	349	296	2	2.46
	1700	281.708	1.31	346	71	36	62	247	201	3	6.71
31/10/94	0800	304.334	0.44	593	96	54	65	377	402	1	1.73
	1700	304.708	0.71	569	92	54	65	376	368	1	2.34

results. The total cations (TZ^+) and anions (TZ^-) are related by the equation, $TZ^+ = 0.98\ TZ^- + 27.6$ with a correlation coefficient of 0.95 for 360 samples. The range of ionic concentrations during the pre-monsoon, monsoon and post-monsoon seasons of the 1994 ablation period is given in Table 3.3. The major anions, sulphate and bicarbonate, range between 85-1140 meq 1^{-1} and 128-1053 meq 1^{-1} respectively. Ca^{2+} was the major cation with a concentration ranging between 236-1941 meq 1^{-1}, and accounts for 66 to 90% of the total cations. K^+ and Mg^{2+} range between 6-20% and 3-15% of the total cations respectively (Table 3.4) shows a comparison of the cationic composition of Dokrianin glacier meltwater with other glaciers.

Table 3.3. Ionic concentration range of Dokriani glacier meltwaters in pre-monsoon, monsoon and post-monsoon season (μeq 1^{-1}), May-October, 1994.

	Discharge m^3s	Ca^{2-}	Mg^{2+}	Na^+	K^+	HCO_3^-	SO_4^{2-}	Cl^-
Pre-monsoon	2.81-0.41	608-262	117-40	65-23	73-45	397-159	418-160	24-2
Monsoon	17.43-1.76	1941-236	80-31	36-12	128-41	1053-128	1140-85	12-1
Post-monsoon	3.92-0.44	593-234	107-37	57-20	68-40	384-168	431-137	7-1

Table 3.4. Summary of cationic composition of Dokriani glacier meltwater in comparison with other glaciers.

Glacier	Period	Year	No. of samples	Range as % (Ca^{2+} + Mg^{2+} + Na^+ + K^+)				Reference
				Ca^{2+}	Mg^{2+}	Na^+	K^+	
1. Dokriani Glacier								
	August-September	1992	180	45-70	3-28	6-18	8-18	
	October-November	1993	88	39-75	9-21	6-18	8-21	
	May-October	1994	360	66-90	3-15	9-15	7-23	
2. Gornergletscher								
	July-August	1974	69	45-90.0	5.2-23.3	1.3-17.4	1.8-22.0	Collins (1979)
	July-September	1975	59	57-71.3	14-18	1.9-8.8	9.6-26.5	Collins (1979)
	July-August	1976	41	52.7-76.1	13.5-17.6	3.2-8.1	3.9-25.0	Collins (1979)
3. Lewis Glacier								
	June-August	1974	14	38.1-39.7	7.5-11.0	44.6-47.0	4.7-12.4	Church (1974)

3.11.1 Sources of Dissolved Ions

The chemical composition of glacial meltwaters has demonstrated high rates of chemical weathering in subglacial environments (Reynolds and Johnson, 1972; Collins, 1979; Raiswell, 1984; Sharp *et al.*, 1995; Brown *et al.*, 1996). The two major anions, HCO_3^- and SO_4^{2-}, in glacial meltwaters are mainly derived from the dissolution of atmospheric carbon dioxide and carbonates, and the oxidation of sulphides (Souchez and Lemmens, 1987). Dissociation of CO_2 and sulphide oxidation are the two main reactions that provide the bulk of the protons to weather carbonates, silicates and aluminosilicate in the drainage basin. The anionic content of glacial meltwater therefore reflects the acid sources used in the chemical weathering of bed rock minerals. The relative proportions of various ions in solution depend on their relative abundance in the bed rock and their solubility (Raiswell, 1984; Tranter *et al.*, 1993). Sources of dissolved ions in the Dokriani glacier meltwaters can be evaluated by studying the relationship among the ions. A parametric correlation analysis for all measured ions during the six month observation period is given in Table 3.5. HCO_3^- and SO_4^{2-} are highly correlated with calcium (r^2, 0.94 and 0.92 respectively) and with potassium (r^2, 0.71 and 0.74). The proportion of HCO_3^- and SO_4^{2-} in the meltwaters reflects the relative dominance of the two major sources of protons during subglacial chemical erosion. According to Brown *et al.*, (1996), the ratio of HCO_3^- to

Table 3.5. Correlation matrix for ions (May-October, 1994)

	Ca^{2+}	Mg^{2+}	Na^+	K^+	HCO_3^-	SO_4^{2-}
Ca^{2+}						
Mg^{2+}	0.50					
Na+	0.35	0.93				
K+	0.76	0.30	0.10			
HCO_3^-	0.94	0.53	0.38	0.71		
SO_4^{2-}	0.92	0.47	0.35	0.74	0.80	
Cl^-	0.15	0.28	0.39	0.14	0.13	0.18

$(HCO_3^- + SO_4^{2-})$, referred to as the C- ratio, has a value of 1.0 when carbonation reactions, involving atmospheric CO_2 dominate [as given in the following equations (1) and (2)] and a value of 0.5, when coupled reactions involving the weathering of carbonates by protons derived from sulphide oxidation dominate [as shown in equation (3)]

$$\underset{\text{(calcite)}}{CaCO_3(s)} + CO_2(aq) + H_2O(aq) = Ca^{2+}(aq) + HCO_3^-(aq) \qquad ...(1)$$

$$\underset{\text{(Ca-feldspar)}}{CaAl_2Si_2O_8(s)} + 2CO_2(aq) + 2H_2O(aq) = Ca^{2+}(aq) + 2HCO_3^- + \underset{\text{(partially weathered feldspar)}}{H_2Al_2Si_2O_8(s)} \qquad ...(2)$$

$$\underset{\text{(pyrite)}}{4FeS_2(s)} + \underset{\text{(calcite)}}{16CaCO_3(s)} + 15O_2(aq) + 14H_2O(aq) =$$

$$16Ca^{2+}(aq) + 16HCO_3^-(aq) + 8SO_4^{2-}(aq) + 4Fe(OH)_3(s) \qquad ...(3)$$

The C-ratio of the Dokriani glacier meltwater during the month of May and June is 0.5 (Figure 3.12), suggesting the dominance of carbonate dissolution and the oxidation of sulphides (Brown *et al.*, 1994). From late June to early July (Julian day 174-187), high discharges were derived largely from snow and icemelt, which rapidly passed through the hydrological system. The enhanced C-ratio (0.55 and 0.7) during this period indicates that protons are additionally derived from the dissolution and dissociation of atmospheric CO_2 (Brown *et al.*, 1996). During the months of July and August (Julian day 195-243), cloud cover over the Himalaya due to the onset of the south-west monsoon, cuts off radiation and causes snowfall in the accumulation zone and rianfall in the ablation region (Collins and Hasnain, 1995). Meltwater discharges during this cloudy monsoonal period were likely to be sustained by drainage of a subglacial reservoir built over the previous months. Hence, runoff çontains a very high proportion of stored subglacial water. The C-ratio during this period varies between 0.31 and 0.63, suggesting the dominance of carbonate dissolution and sulphide oxidation in controlling the meltwater composition. The cloud burst on August 24 (Julian day 236)[1] resulted in high discharge (17.45 m^3 s^{-1}) and the lowest C-ratio (0.31), suggesting that rainfall passing through supraglacial moraine may also have a strong control on the chemical characteristics of meltwater.

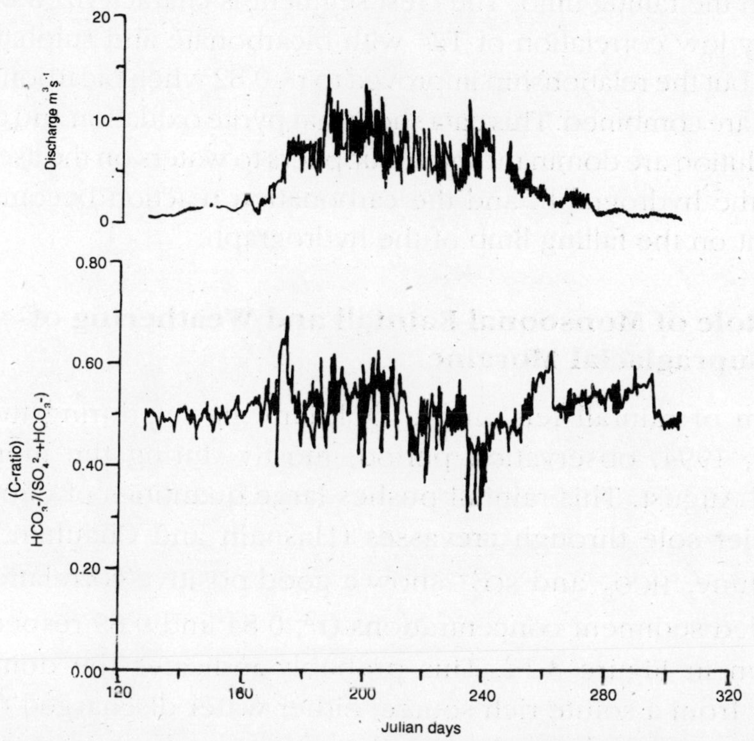

Fig. 3.12. Ratios of HCO_3^- to (HCO_3^- + SO_4^{2-}) in portal meltwaters draining the Dokriani glacier, Ganga headwater, Garhwal Himalaya. Reprinted from Hasnain and Thayyen (1999a)

---

1 Julian day. The number of days which have elapsed since 12.00 GMT on 1 January. This consecutive numbering of days gives a calender independent of month and year which is used for analysing periodic phenomena. This system was devised by Julian Scaliger. The formula used to find out Julian day is as follows:

$$\text{Julian day} + \left(\frac{h}{24}\right) \text{ for eg. 23-06-1994} = 296 + \left(\frac{h}{24}\right).$$

Figure 3.13 shows the relationship between SO_4^{2-} and HCO_3^- to total cations for the rising limb, the crest segment and the falling limb of all hydrographs for May through October. Both sulphate and bicarbonate shows good correlation with TZ^+ on the rising limb, with sulphate showing better correlation, whereas bicarbonate was better correlated to TZ^+ on the falling limb. The crest segment is characterized by comparatively low correlation of TZ^+ with bicarbonate and sulphate individually, but the relationship improved to r^2, 0.82 when bicarbonate and sulphate are combined. This data show that pyrite oxidation and carbonate dissolution are dominant solute suppliers to waters on the ascending limb of the hydrograph, and the carbonation reaction become more important on the falling limb of the hydrograph.

3.11.2 Role of Monsoonal Rainfall and Weathering of Supraglacial Moraine

1269 mm of rainfall fell on the Dokriani glacier during the May-October, 1994, observation period, mostly during the months of July and August. This rainfall pushes large quantities of debris into the glacier sole through crevasses (Hasnain and Chauhan, 1993). During June, HCO_3^- and SO_4^{2-} show a good positive correlation with suspended sediment concentrations (r^2, 0.84 and 0.89 respectively) as shown in Figure 3.14. This probably indicates the dominance of water from a solute rich source, either water discharged through sediments at the glacier bed or that has been kept in storage for a long period, as a major transporter of suspended sediment. Successive diurnal melting of snow and ice rapidly leaches most of the sulphate from the snow and ice (Davies *et al.*, 1982). Hence, later in the ablation season, the leached snow and ice give rise to sulphate depleted supraglacial meltwater (Tranter and Raiswell, 1991). The release of sulphate by the oxidation of sulphides in suspended sediment is a slow process and therefore is only likely to contribute significant sulphate concentrations to waters which have spent long periods at the glacier bed in contact with sedi-

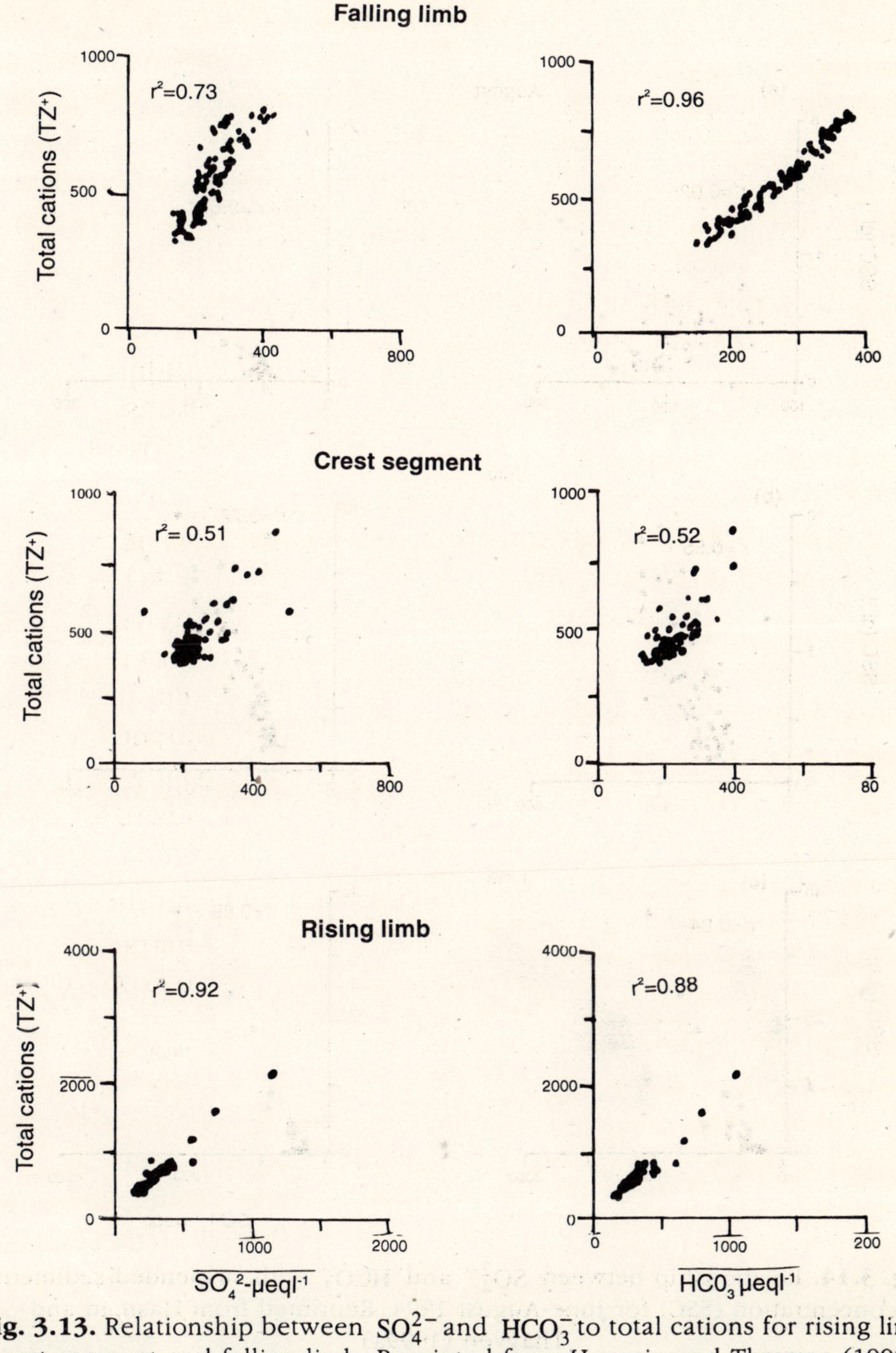

Fig. 3.13. Relationship between SO_4^{2-} and HCO_3^- to total cations for rising limb, crest segment and falling limb. Reprinted from Hasnain and Thayyen (1999a).

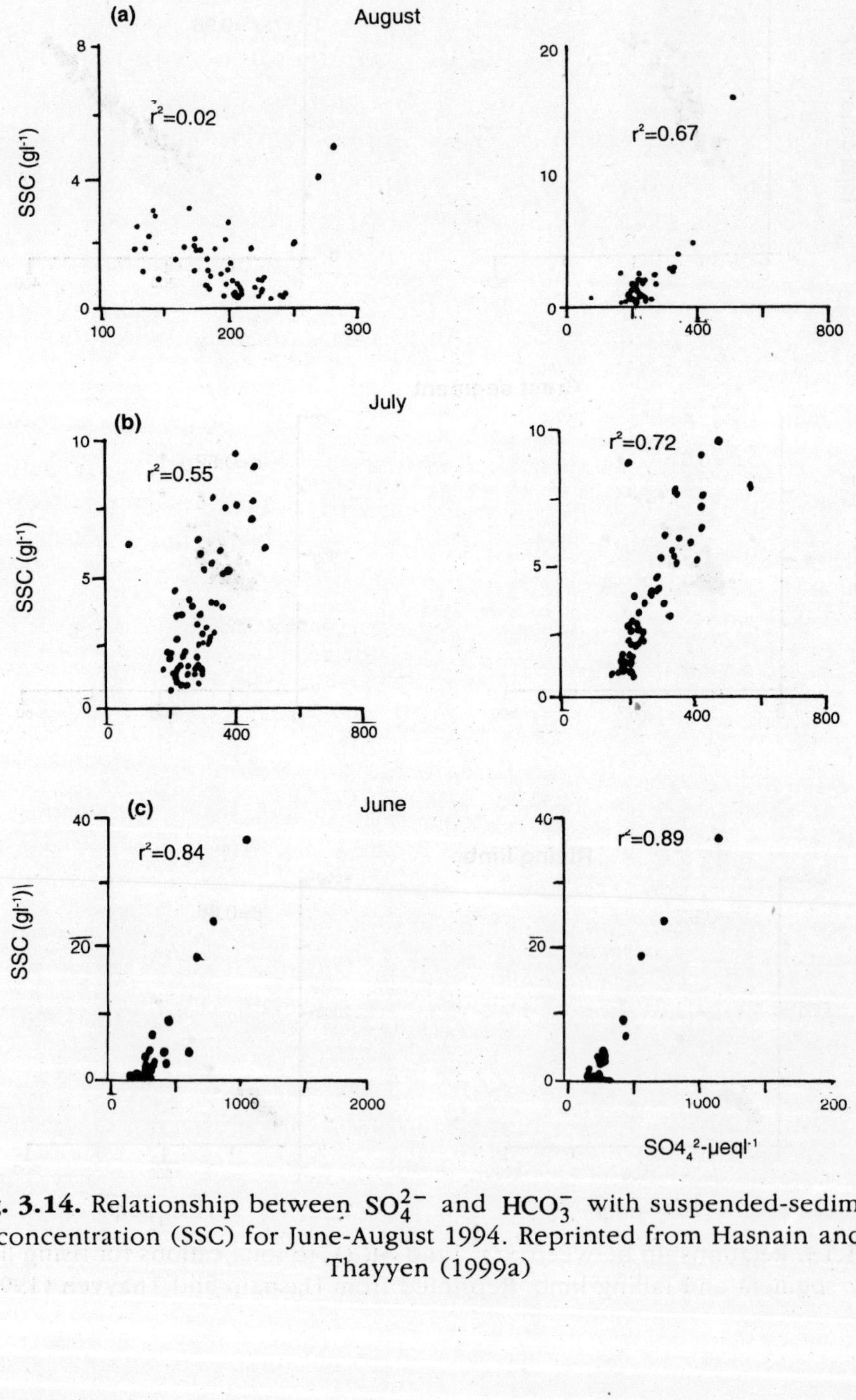

Fig. 3.14. Relationship between SO_4^{2-} and HCO_3^- with suspended-sediment concentration (SSC) for June-August 1994. Reprinted from Hasnain and Thayyen (1999a)

ments (Brown *et al.* 1994). However, contrary to these observations, Dokriani glacier meltwaters show higher sulphate concentrations during July and August and also show good positive correlation between SO_4^{2-} and suspended sediment concentration (r^2, 0.72 and 0.67 respectively) as shown in Figure 3.14. One possible explanation follows: continued low intensity rainfall may have initiated pyrite oxidation within the supraglacial moraines. Therefore, occasional high intensity showers may leach the sulphate from the debris and cause its dominations during the seasonal high flow period of July and August, when suspended sediment concentration (SSC) are high. The percentage contribution of suspended sediments from the supraglacial sources in the portal flow is not known. The post-monsoon dry period of September exhibits a weak correlation between SO_4^{2-} and suspended sediment concentration ($r^2 = 0.09$) and thus substantiate the observation of Tranter *et al.* (1993) that reactive sulphides will be largely absent from rock flour transported from the subglacial system.

3.11.3. Chemical Denudation Rates

The cationic denudation rates from some of the important glaciers in the European Alps and North America are given in Table 3.6. The cationic denudation rate of Dokriani glacier is 4160 meq m^{-2} yr^{-1}. It is about four times higher than the highest recorded at other glaciers, namely Gornergletscher, Swiss Alps (Metcalf, 1986). The Chemical Denudation Rate (CDR) of Dokriani glacier was 321 tons km^2 yr^{-1} Monthly values of chemical denudation during the 1994 observation period are given in Table 3.7. The high chemical denudation rate substantiate Hasnain and Thayyen (1999a) observation that the high relief, south-west monsoonal rainfall and supraglacial moraines are very significant in controlling the chemistry of Garhwal Himalaya glacial meltwaters.

Table 3.6. Cationic denudation rates of Dokriani glacier in comparison with other glaciers

No.	Name of the glacier	Denudation rates (meq m^{-2} yr^{-1})	Reference
	(European Alps Glaciers)		
1.	Haut glacier d'Arolla (Swiss Alps)	640-685	(Sharp *et al.*, 1995)
2.	Glacier de Tsijiore Nouve (Swiss Alps)	508	(Souchez & Lemmens, 1987)
3.	Gornergletscher (Swiss Alps)	454	(Collins, 1983)
4.	Gornergletscher (Swiss Alps)	1101 ± 505	(Metcalf, 1986)
	(North American Glaciers)		
5.	South Cascade glacier (Washington State, U.S.A.	930	(Reynolds & Johnson, 1972)
6.	Beremdon glacier (Canada)	947	(Eyles *et al.*, 1982)
	(Himalaya)		
7.	Dokriani glacier	4160	(Hasnain & Thayyen 1999)

Table 3.7. Monthly solute flux and chemical denudation rates of Dokriani glacier.

Month	Solute flux (tons)	Solute flux %	Denudation rate (tons/km^2)
May	86	3	9.0
June	430	18	45
July	970	40	100
August	560	23	59
September	280	11	29
October	120	5	13

4

HYDROGRAPH SEPARATION

4.1 INTRODUCTION

Glacial meltwaters exhibit diurnal, inverse relationships between discharge and total dissolved solids (or conductivity) during the ablation season at ambient water temperature of 0-2°C as shown in Figure 4.1. In general, the maximum conductivity coincided with minimum dis-

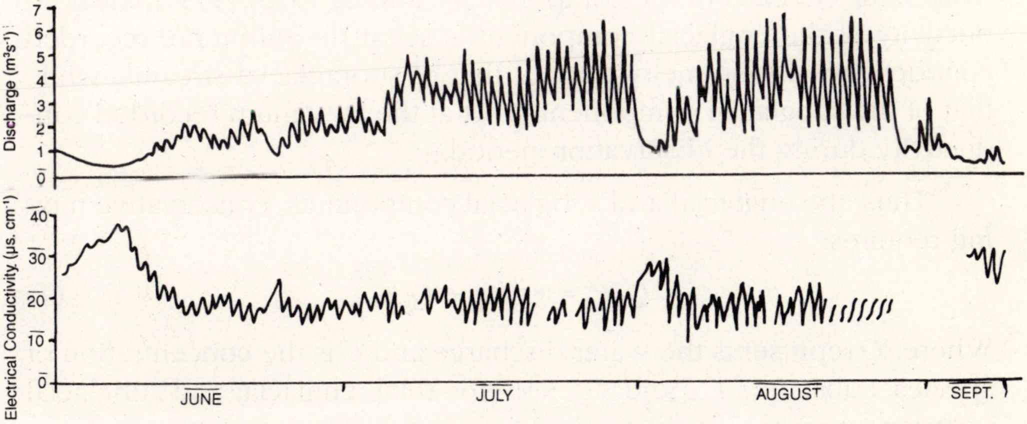

Fig. 4.1. Time series of hourly discharge and electrical conductivity at ambient water temperature of –2 °C of Swiss Alps glacier. Reprinted from Gurnell *et al.* (1994)

charge during each diurnal cycle, but the minimum conductivity often preceded maximum discharge by one or two hours (Gurnell *et al.*, 1994). Early in the season diurnal cycles were absent from the time series, but after mid-June subdued cycles had commenced in both series. However, during July and August, minimum discharge and maximum conductivity occurred between 800 and 1000 h; maximum discharge occurred between 1600 and 2000 h. This provided a good match with the times of solute sampling and thus ensured that the twice-daily sampling was likely to be close to the extremes in solute concentrations.

Each day, maximum concentration of solute is found in meltwaters at low discharge, and minimum concentration at maximum discharge. Therefore, total hydrograph can be separated, based on ionic concentrations, into englacial and subglacial components. The contact of englacial component generated by the surface melting and passing through the ice-walled channels, in which the reactive materials are little or absent and hence the potential for solute acquisition is limited and waters are relatively dilute. By contrast, the subglacial component, relatively has slow transit through the debris-rich basal conduits. The subglacial water are thus relatively concentrated because of longer contact times.

Hydrograph separation has been performed by assigning conductivity values to each of the components (Collins, 1978, 1979). The conductivity of the englacial component is set at the minimum recorded conductivity of total meltwaters or that of supraglacial streams, while that of the subglacial component is set at the maximum recorded conductivity during the observation period.

Thus, the englacial and subglacial components, conservative mixing requires:

$$Q_t C_t = Q_e C_e + Q_s C_s \qquad (1)$$

Where, Q represents the water discharge and C is the concentration of species. Subscripts t, e and s denote the total, englacial and subglacial components respectively.

Continuity of flow requires that

$$Q_t = Q_e + Q_s \tag{2}$$

Combining Equations (1) and (2) gives the following equation.

$$Q_e = Q_t\,(C_s - C_t)/(C_s - C_e) \tag{3}$$

Equation (3) illustrates that four parameters must be determined to perform a hydrograph separation; of these, only two, Q_t and C_b are routinely measured. Hence the other two parameters, C_e and C_s need to be evaluated by independent means. The usual method is to equate C_e and C_s with the minimum and maximum conductivity values respectively recorded during the sampling season (Collins, 1979).

Tranter and Raiswell (1991) have suggested a method of separating the hydrograph into englacial and subglacial components by using sulphate as a marker of subglacial component.

4.2 ELECTRICAL CONDUCTIVITY MEASUREMENTS

Electrical conductivity measurements may be made electrometrically (i.e. by immersing a conductivity cell within the fluid to be measured) with apparent ease and accuracy, continuous records may be obtained by transferring the signal from a cell fixed within the flow to a data logger. Meltwater conductivity's, continuous or discrete, have a variety of applications in glacier hydrology and glacier hydrochemistry, as e.g.

(1) To indicate the nature and extent of temporal variations in meltwater quality

(2) To indicate the nature and extent spatial variations in meltwater quality

(3) To provide quantitative estimate of solute load

(4) To facilitate separation of hydrochemically distinct contributions to total meltwater runoff

(5) To identify solute sources and pathways

(6) To provide a check on measured specific ion concentrations.

4.2.1 Estimate of Dissolved Ion Discharge Using Year-round Electrical Conductivity Measurements

Collins (1981) has pioneered the *in situ* conductivity measurement of glacial meltwaters throughout the year. He used conductivity as a surrogate measure of total dissolved solids concentration in meltwater to describe the seasonal variation of solute in a semi-quantitative manner. He concluded that almost all the solute load beneath glaciers is removed during ablation period. He also reasoned that since meltwater probably does not have access to all areas of the bed, meltwater hydrochemically probably underestimates total chemical activity. However, a linear correlation between total cations and electrical conductivity may be established. The cation load is multiplied by 2 to estimate the total dissolved ion load, by charge balance considerations. The total dissolved ion discharge is then estimated by numerical integration of the total ion concentration times water discharge over a finite time interval, through an annual cycle, by using the following equation:

$$F = \sum_{i=1}^{n} \left[\frac{2(fK_i - b)}{m} \frac{Q_i \Delta t_i}{A} \right] \quad (4)$$

where:

F = the chemical erosion of ionic solutes during a time interval of $\sum_{i=1}^{n} \Delta t_i$ over the area of the catchment

K_i = the mean electrical conductivity at a stream temperature of 0-1°C during the time interval i,

f = the increase of electrical conductivity from *in situ* temperature to the temperature of cation determination.

b, m = constants established by least squares determination of Σ cations versus conductivity in the laboratory from:

$$Y = mX + b \quad (5)$$

where:

X = Σ cations and Y = lab conductivity;

$\frac{(fK_i - b)}{m}$ = the estimate of the mean cation sum during the time interval *i*,

$\frac{2(fK_i - b)}{m}$ = the estimate of the total ions (mean value) in solution,

Q_i = the mean stream discharge during the time interval *i*

Δt_i = the time interval of convenience

and A = the catchment drainage area.

It is important to distinguish total dissolved ions from total dissolved solids because such important species as dissolved silica are usually not dissociated significantly at the pH of glacial waters; hence, they have no effect on, or good correlation with, conductivity. The field data from Gornergletscher catchment, Switzerland is used here as an example to estimate dissolved ion discharge. Table 4.1 gives the summation of four major cations range from 28-4650 μS cm^{-1} at 18 to 19°C. The range of *in situ* conductivity values for the Gornergletscher catchment was 8 to 115 μS cm^{-1} at 0-2°C.

Table 4.1. Determinations of lab conductivity and major cations for water samples from the Gornergletscher catchment, Switzerland

Sample No.	$\Sigma Ca^{2+} + Mg^{2+} + Na^{+} + K^{+}$ (μEl^{-1})	Lab conductivity (μScm^{-1} at 18-19°C)
1	240	17.6
2	4580	213
3	4650	214
4	34	7.4
5	28	3.3
6	56	3.7
7	30	2.9
8	30	2.9
9	34	3.4
10	320	25.6
11	480	30.0

Table 4.1. (Contd.)

Table 4.1. *(Contd.)*

Sample No.	$\Sigma Ca^{2+} + Mg^{2+} + Na^{+} + K^{+}$ (μEl^{-1})	Lab conductivity (μScm^{-1} at 18-19°C)
12	300	18.0
13	860	43.2
14	93	6.9
15	75	6.1
16	180	1.5
17	44	3.4
18	74	4.4
19	370	18.7
20	400	22.0
21	280	17.2
22	150	9.4
23	2200	131
24	29	2.2
25	810	47.8
26	910	50.2
27	920	48.6
28	900	43.2
29	320	18.1
30	290	18.0
31	350	20.9
32	280	15.0
33	260	15.0

The least squares analysis of the data in Table 4.2 according to equation 2 results in:

$$Y = 0.0469\ X + 3.616 \qquad (6)$$

where: Y is in $\mu S\ cm^{-1}$ at 18-19° C and X is the major cations sum in μ E L^{-1}. The correlation coefficient is 0.994 and $\tau_n(Y)$ = 51.9 and $\tau_n(X)$ = 1099.4. For use in calculating the dissolved ion load from conductivity, Equation(3) can be rearranged to :

$$X = \frac{Y - 3.616}{0.0469} \qquad (7)$$

This equation predicts most of the cation sums in Table 4.1 within 10%.

Table 4.2. Dissolved ion discharge for maximum conductivity, Gornergletscher, Switzerland

Time of year	Q_1 (m^3s^{-1})	K_1 ($mScm^{-1}$)	$1.60K_1$ ($mScm^{-1}$)	$\frac{(fK_1-b)}{m}$ (mEL^{-1})	$\frac{2(fK_1-b)}{m}$ Q_1(mEL^{-1}) m^3s^{-1})	Dt_1(mon.)	FA($mEL^{-1}m^3$) (Dt_1^{-1})
Winter (Dec. - Apr.)	0.1-0.2	90-115	144-184	2993-3845	599-1538	5	7.76×10^9 1.99×10^{10}
Spring (May)	0.2-5.7	30-89	48-142	946-2950	378-33630	1	9.80×10^8 8.72×10^{10}
Summer (June-Sept.)	8.3-20.2	10-41	16-66	264-1330	4382-53732	4	4.54×10^{10} 5.57×10^{11}
Autumn (Oct.-Nov.)	0.2-0.8	40-51	64-82	1287-1671	515-2674	2	2.67×10^9 1.39×10^{10}
						12	

Without a detailed understanding of meltwater chemistry changes upon heating, it is currently impossible to give a simple temperature correction for the conductivity of glacial meltwaters. However, it is possible to state minimum and maximum values as limiting boundary conditions for such changes. The minimum conductivity change would be zero during heating from 0-2°C to 18-19°C. The maximum observed increase in meltwater samples was 60%. For example , the minimum conductivity gives an ionic denudation rate for the 82 km^2 Gornerglestcher catchment as 460-5180 mE m^{-2} a^{-1}. The maximum conductivity gives a range of 730-8290 mEm $^{-2}$ a^{-1}. Reynolds and Johnson (1972) calculated the ionic denudation rate of South Cascade glacier to be 9300 m^{-2} a^{-1}, significantly greater than world average rate of 390 mE m^{-2} a^{-1}. Their estimate based on summer months sampling and Collins (1981) also observed that most chemical evacuation takes place during summer times in an Alpine glacier system. Thus, Reynolds and Johnson (1972) claim that some glacial catchments have significantly greater ionic denudation rates than the world average has been verified.

4.2.2 CO_2 Effects

Meltwaters emerging from inter-glacial conduits will be undersaturated with respect to atmospheric CO_2, and may rapidly dissolve CO_2 in the proglacial zone. Reynold and Jonhson (1972) also commented that the turbulent, will aerated flow conditions characteristic of meltwater environments may promote chemical weathering by maintaining the acid potential of the water.

Glacier-fed streams transport high and variable quantities of sediment in suspension (Ostrem, 1975; Collins, 1979c; Fenn *et al.*, 1985). It was contended by Collins (1979) that neither component exhibits any temporal variation in chemical composition and that no post-mixing chemical evolution of meltwaters occurs. However it is likely that each component will demonstrate some temporal variation during the course of the ablation season (e.g. due to the development of different hydrological routing and that post-mixing chemical evolution will occur when relatively reactive sediment is transported into and through the mixing environment (Raiswell and Thomas, 1984, Tranter *et al.*, 1989).

4.2.3 EC-Discharge relationship

Electrical conductivity (EC) and discharge variations for the Dokriani glacier, Garhwal Himalaya, is shown in Figure 4.2. The inverse relationship between electrical conductivity and discharge is clear, as are the diurnal cycles in both variables throughout the ablation season. In general, the maximum conductivity coincided with minimum discharge during each diurnal cycle, but the minimum conductivity often preceded maximum discharge by one or two hours. Early in the season diurnal cycles were absent from time series, but after mid-June subdued cycles are distinctly visible. The sharp diurnal peaks, however, in July and August were caused by the impact of monsoonal rainfall. However, the EC-Q scatter plot (Figure 4.3.) for the Dokriani glacier meltwaters invariably indicates an inverse relationship between the two variables which is often non-linear and shows a widely varying degree of scatter. The degree of scatter differs from basin to basin (and from time to time in a given basin) according to the extent of variability in the quantity-quality characteristics of runoff components, and also according to the extent hysteresis effects generated by diurnal and longer period lags between the Q and EC series (Fenn, 1987). The ablation season scatter plot as shown in Figure 3.10 in Chapter 3 for the Gornera glacier, Switzerland, (Collins, 1979a) illustrates the trapezoidal relationship. It also indicates an inverse, non-linear relationship between the two variables with widely varying degree of scatter.

4.2.4 Supraglacial EC Studies

Supraglacial EC studies on Dokriani glacier, Garhwal Himalaya, India were carried out in November 1993. The results were reported by Hasnain and Thayyen, (1996). The samples were collected from different locations of ablation zone and shows remarkable spatial variations as evident from Figure 4.4. Supraglacial EC values range from 0.6 μS cm^{-1} to 3.7 μS cm^{-1}. A large part of the ablation zone of Dokriani glacier is covered by lateral surface moraines leaving the middle part more or less debris free. The samples collected from the debris covered area show higher EC values. The ice mass under the debris cover obviously holds more sediments and shows higher EC values after melting. Fenn

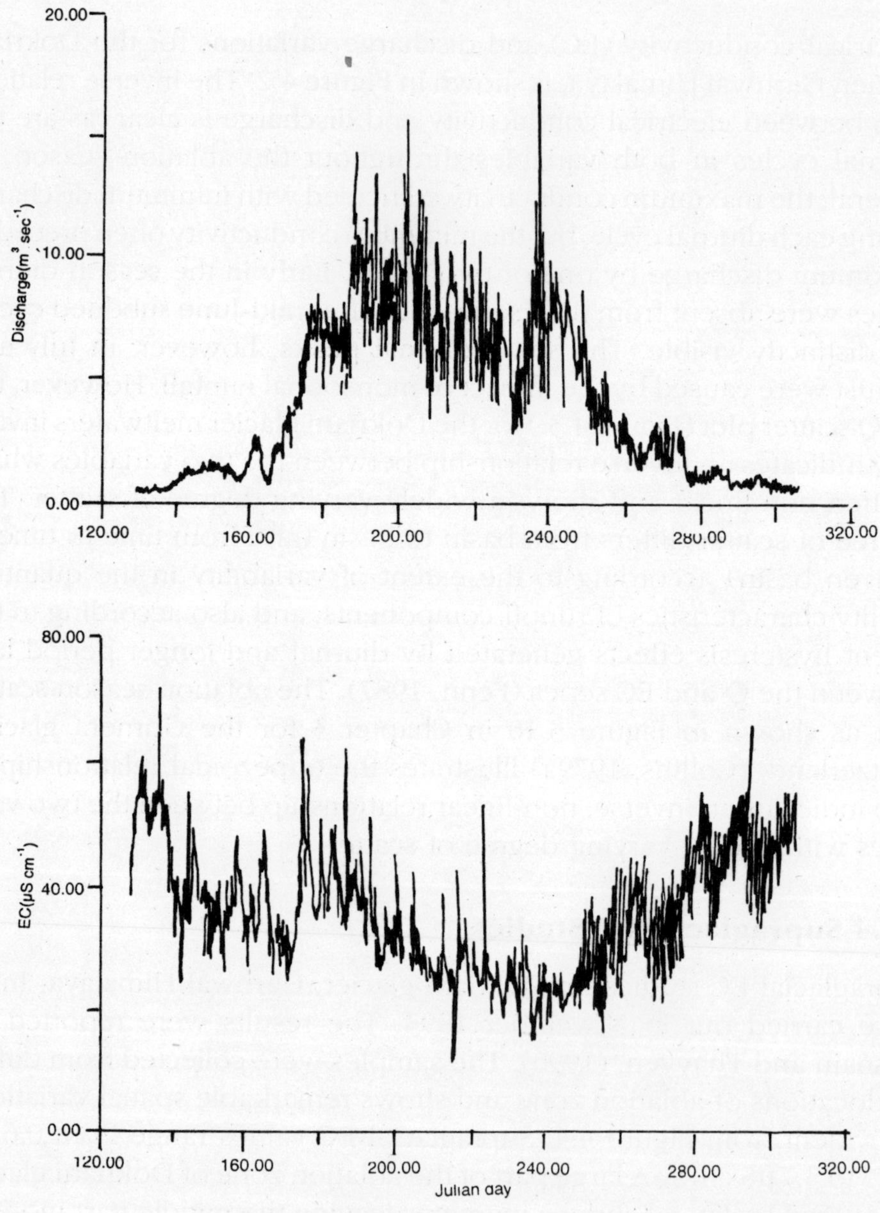

Fig. 4.2. Time series of hourly discharge and electrical conductivity at ambient water temperature of 0-2°C of Himalayan glacier. Reprinted from Hasnain and Thayyen (1996)

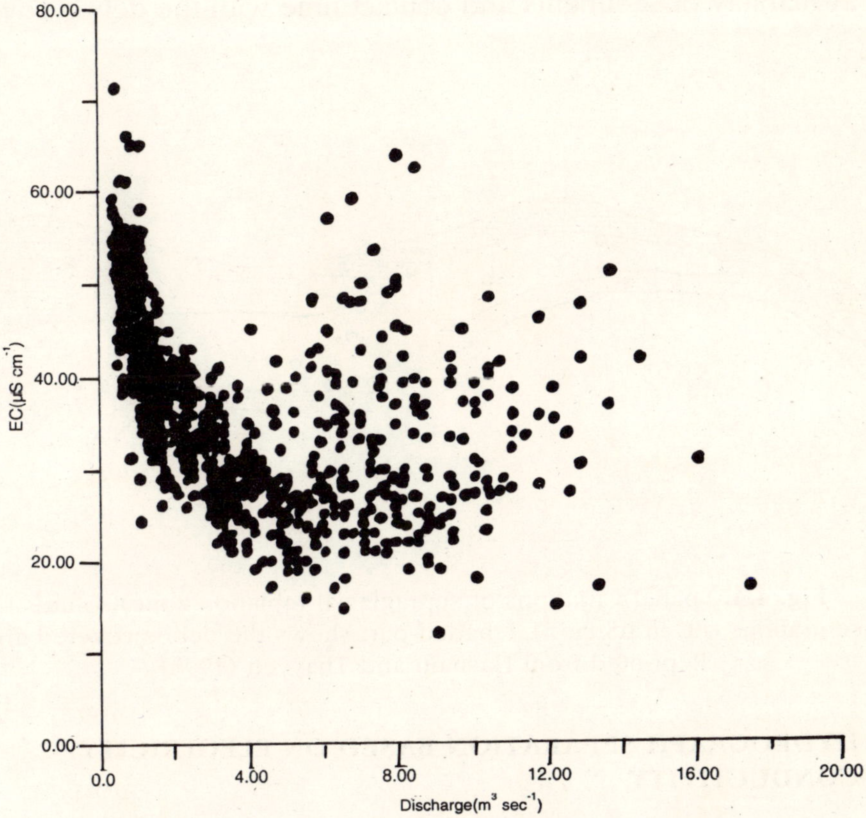

Fig. 4.3. Scatter plot of hourly discharge and electrical conductivity. Reprinted from Hasnain and Thayyen (1996)

(1987) also pointed out that the concentration of supraglacial meltwaters draining from the zones of surface maraine would be higher. EC values of supraglacial meltwater channels show an average of 1.6 μS cm^{-1}. EC values of meltwaters of various supraglacial environment show distinct variations. In the year 1992, EC values are higher than those in 1993 and range from 3.6 μS cm^{-1} to 4.2 μS cm^{-1} due to extended travel time through debris-covered glacial surface. Water stored in fissures along with higher concentration of sediment shows EC value of 6.2 μS cm^{-1} and supraglacial stream flowing in contact with the surface moraine shows the highest value of 10.6 μS cm^{-1}. It is clear from these observations that the ionic enrichment of supraglacial meltwaters depends on the availability of sediments and contact time with the debris cover.

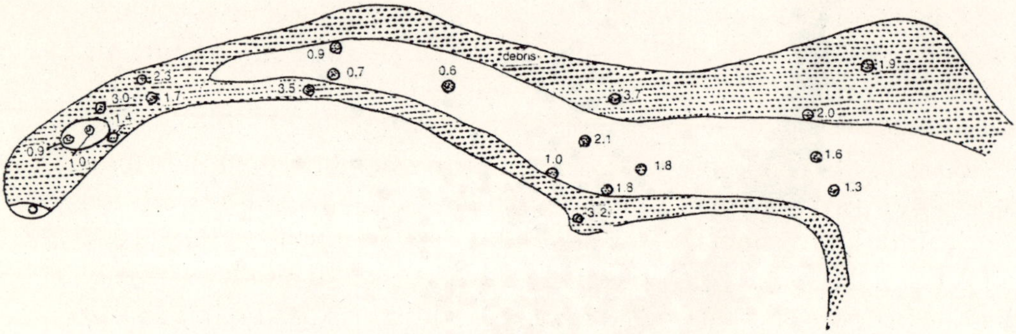

Fig. 4.4. Spatial variations of supraglacial (ablation zone) solute concentrations (EC in μS cm^{-1}). (shaded part shows the debris covered area). Reprinted from Hasnain and Thayyen (1996)

4.3 HYDROGRAPH SEPARATION BASED ON ELECTRICITY CONDUCTIVITY

The subglacial component of total meltwater emerging from the portal of Dokriani glacier, Garhwal Himalaya can be separated by using the following expression:

$$Q_t = [(EC_t - EC_e)/(EC_s - EC_e)]\ Q_t$$

The values of parameters Q_t, EC_t and EC_e are required to perform the hydrograph separation. Q_t and EC_t can be determined directly from field observations. The supraglacial meltwaters considered to be passing through the englacial system are without ionic enrichment. Spatial as well as temporal variations in electrical conductivity of supraglacial meltwaters were observed during the sampling periods. These variations are related to the contact time of supraglacial meltwaters with the debris cover. Hence average *EC* values of supraglacial meltwaters were taken as the representative EC_e values. Seasonal variations in solute characteristics of supraglacial meltwaters are related to the travel times through the debris-covered areas and the amount of water derived from the surface moraine zones. Therefore different values were selected to represent the englacial component of *EC* during 1992 and 1993 observation periods. During the recession flow periods, the diurnal cycle is suppressed and the chemically enriched waters released from subglacial or groundwater storage constitute most of the total flow. The *EC* of meltwater emerging from the glacial portal may often approach the *EC* of subglacial waters at such times. Flushing out of chemically enriched water from subglacial storage's is a usual phenomenon with the glacial discharges and these events represent the unusual *EC* peak events in the chemograph. The *EC*s values were selected from recession flow regimes and neglected the unusual *EC* peak events observed during the sampling periods. The selected values of EC_e and EC_s in µS cm^{-1} are as follows:

	1992	**1993**
EC_e	4	1.8
EC_s	56	54

The separated englacial and subglacial components of total discharge of Dokriani glacier during 28 August to 20 September 1992, are shown in Figure 4.5. The englacial flow component shows inphase rhythmic variations with the net discharge and shows steadily decreasing trend along with the decreasing trend in total discharge during the observation period. Surface meltwaters available for transportation rapidly flow through the englacial conduit system and control the characteristics of total flow hydrograph. The subglacial component shows low amplitude diurnal variations during 28 August to 9 September when 64 % of the total flow was contributed by the englacial system and within the range of 2.21 m^3 s^{-1} to 0.53 m^3 s^{-1}. Perhaps, reduced surface melting during 10 to 20 September due to lowering of atmospheric temperature resulted in reduction of net flow with 62.5% subglacial contribution. During this period subglacial component shows nominal increasing trend along with the decreasing trend in total and englacial discharges. The cooling of climate leads to the closure of near-surface englacial channels and diverts the available surface meltwater into deeper channels which may be routed through solute-rich environment and contributes to the sub-glacial components. The hydrograph of the flow through the englacial component exhibits multiple peaks which are proportional to the complex hydrograph of the total flow and confirms that the precipitation during the period has little effect on net flow hydrograph.

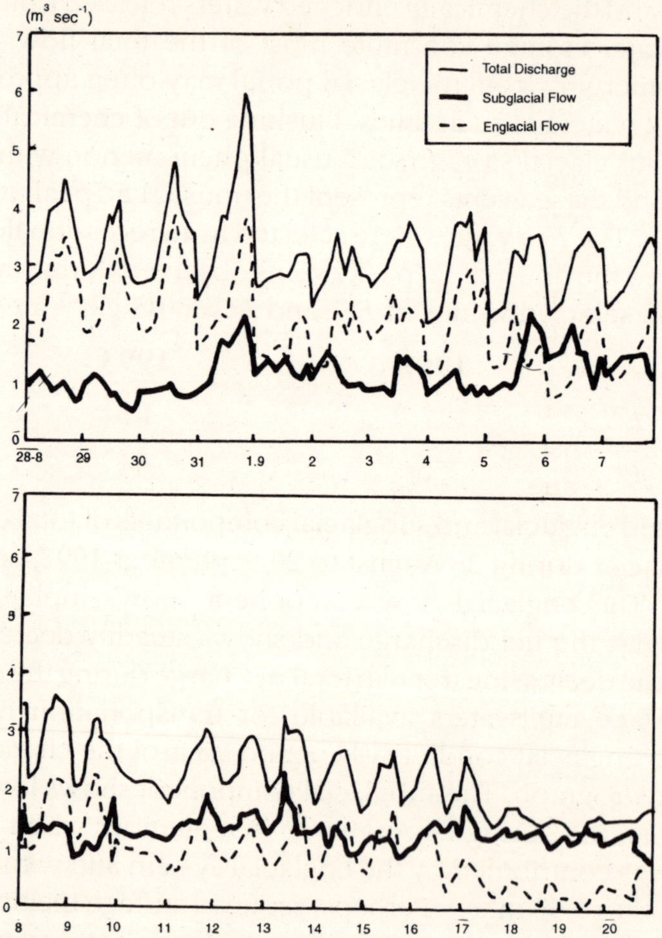

Fig. 4.5. Separated subglacial and englacial flow components of portal discharge of Dokriani glacier, 28 August to 20 September 1992. Reprinted from Hasnain and Thayyen (1994)

Separated components of flow from 23 October to 17 November 1993 are shown in Figure 4.6. During the observation period, subglacial

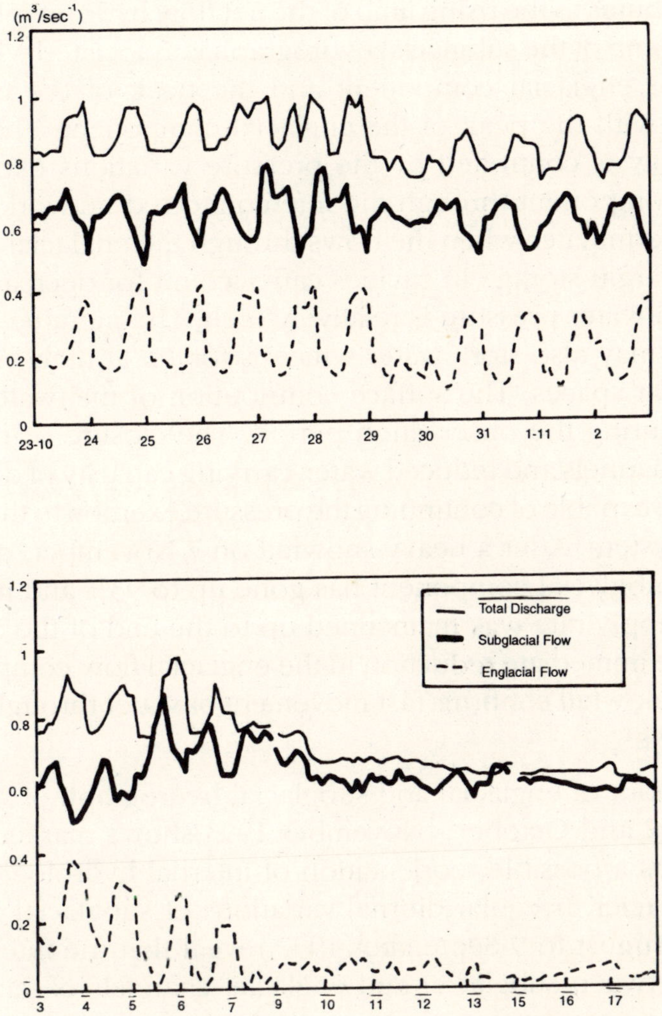

Fig 4.6 Separated subglacial and englacial flow components of portal discharges of Dokriani glacier. 23 October to 17 November 1993. Reprinted from Hasnain and Thayyen (1994).

contribution was much higher than the englacial component. From 23 October to 6 November subglacial flow contributed 74.7 % of the total flow. During this period both englacial and subglacial components show diurnal variations and the peak of the englacial component lagged by

3-5 h to the peak of the subglacial component. The subglacial component contributes to the rising limb of the net flow hydrograph. Generally the falling limb of the subglacial hydrograph is associated with the rising limb of the englacial component and the peak of the total flow is associated with the peak of the englacial component. The subglacial channel flow is controlled by the pressure variations exerted by the englacial component through the interconnected network. Subglacial discharge terminates when the flows through the englacial system rise. Channel margin storage in cavities can account for decreased conduit flow, when water pressure is relatively high. The moraine layer under the glacier may also store water which is forced at high pressure into intergranular spaces. The surface contribution of meltwaters was less available during the observation period. The closure of near-surface englacial channels and reduced water-carrying capacity of deeper channels may be capable of continuing the pressure exertion to the subglacial channels system. After a heavy snowfall on 7 November, contribution from the subglacial component has gone up to 93% and more or less constant supply rate was maintained up to the end of the observation period. The immediate reduction in the englacial flow component after the heavy snowfall confirms fast movement of water through the englacial channels.

Behaviour of englacial and subglacial hydrograph of August-September 1992 and October - November 1993 shows marked variations and suggests a possible reorientation of internal hydrologic system of Dokriani glacier. Irregular diurnal variations of subglacial component during 28 August to 9 September 1992 reveal that the interconnected network between subglacial and englacial channels of varying travel time was more active than those of the 1993 observation period.

The field EC measurements in two consecutive years (1992, 1993) have been effectively used to separate the components of bulk flow emanating from the Dokriani glacier. The study provides as insight into the internal hydrologic system of the Himalayan glacier. The study further concludes that there exists an interconnected hydrologic channels network and this perhaps is causing the diurnal fluctuations in the subglacial component.

4.4 HYDROGRAPH SEPARATION BY USING SULPHATE AS A MARKER

Hydrograph separation of the englacial and subglacial components can be performed by sulphate concentration (Tranter and Raiswell, 1991). Successive diurnal melting of snow and ice rapidly leaches most of the sulphate from the snow and ice crystals (Davies *et al.*, 1982). Hence, later on in the season, the leached snow and ice gives rise to sulphate depleted supraglacial meltwaters. Since supraglacial meltwaters evolve to form waters of the englacial component, the sulphate concentration of the englacial component ought to be negligible during the early melt season.

The Gornergletscher, Switzerland data given in Table 4.3 clearly indicate that the total sulphate concentration is close to zero at maximum discharge. Since the englacial component dominants at maximum discharge, it follows that the sulphate concentration of the englacial component must also be negligible. By this reasoning, Tranter and Raiswell (1991) assumed that the sulphate concentration of the englacial component is zero. Hence, sulphate is only found in the subglacial component, where it is formed by the oxidation of pyrite (Lerman, 1979). Pyrite is commonly found in many sedimentary, metamorphic and igneous rocks. Therefore, sulphate is formed from the oxidation of pyrite throughout the subglacial hydrological system where freshly comminuted rock flour comes into contact with oxygenated meltwaters.

Table 4.3 The range in composition of supraglacial meltwaters from Gorner glacier, Swiss Alps. (after Tranter and Raiswell, 1991).

Ca^{2+}	14-44
Mg^{2+}	0-17
K^+	0-4
Na^+	0-3
HCO_3^-	36-72
SO_4^{2-}	< 0.01

Since sulphate is not found in significant concentrations in the englacial component, it acts as a marker for the subglacial component. Linear relationship between total sulphate and discharge has also been observed in other Alpine glaciers, Les Bossons, French Alps (Tranter, 1982), upper Arolla glacier, Swiss Alps (Brown and Tranter, 1990).

Two component mixing relationships are normally plotted with the independent variable representing the mass fraction of one of the components. The composition of each component or end member is defined when the mass fraction is 0 and 1, respectively.

For sulphate, assuming $C_e = 0$; Equation (3) can be rewritten as:

$$Q_c/Q_b = (S_e - C_b)/C_s \quad (8)$$

where the superscript *s* denotes sulphate. There is a linear relationship between total sulphate and discharge. Hence,

$$C_b = mQ_b + c \quad (9)$$

where *m* and *c* are constants.

Hence, combining Equations (8) and (9) gives

$$Q_e/Q_b = 1\text{-}(c/C_s) - (m/C_s)\, Q_b \quad (10)$$

On the righthand side of Equation (10), only Q_b is a variable. Hence, the bulk discharge, Q_b, is a simple measure of the mass fraction of the englacial component in the total discharge. The subglacial component is defined when the mass fraction of the englacial component is zero, that is, when the lefthand side of Equation (10) is zero. Examination of Equation (10) reveals that the maximum concentration of sulphate in the subglacial component is c, since by setting the englacial component to zero, C_s is maximum when Q_b is zero.

It has been generally observed that the waters of the delayed flow on the diurnal cycle contain both high concentrations of HCO_3^- and SO_4^{2-}. Figure 4.7 illustrates that the mass fraction of SO_4^{2-} (defined as the ratio of sulphate to the sum of sulphate and bicarbonate, where concentrations are in units of equivalents) in the portal meltwaters draining Haut Glacier d' Arolla decreases as discharge increases. This suggests that delayed flow, which dominates at lower discharge, contains a higher mass fraction of sulphate than quickflow (Tranter *et al*., 1993). This type of relationship, however, is not found valid in the meltwaters draining from the Dokriani glacier, Garhwal Himalaya, India as shown in Figure 4.8. Perhaps the quick flow in the Himalayan glaciers is always

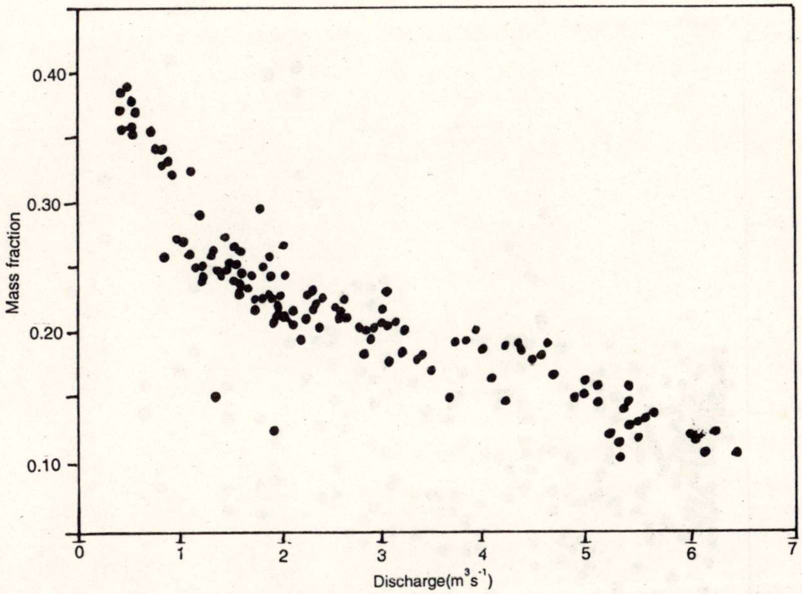

Fig. 4.7. The mass fraction of $SO_4^{2-}(SO_4^{2-}/SO_4^{2-}) + HCO_3^-)$ and where units are equivalents, as a function of discharge for portal meltwaters draining Haut Glacier d' Arolla throughout the 1989 ablation season. Reprinted from Tranter *et al.* (1993).

interacting more with extensive moranic debris in the supraglacial and englacial channels and in the process got enriched with sulphate.

Probably continuous low intensity rainfall may have initiated pyrite oxidation within the supraglacial moraines. Therefore, July and August monsoonal showers may leach the sulphate from debris and cause its domination during the seasonal high flow period of when suspended sediment concentrations are high as a result of debris avalanching. The percentage contribution of suspended sediments from the supraglacial sources to the portal flow in not yet quantified. This suggests that the monsoonal rainfall enhances the supraglacial moraine weathering, providing high sulphate content in the meltwater during periods of high discharge. Consequently, sulphate is not suitable to use as a marker to separate flow components in the meltwater of the southern Himalayan glaciers.

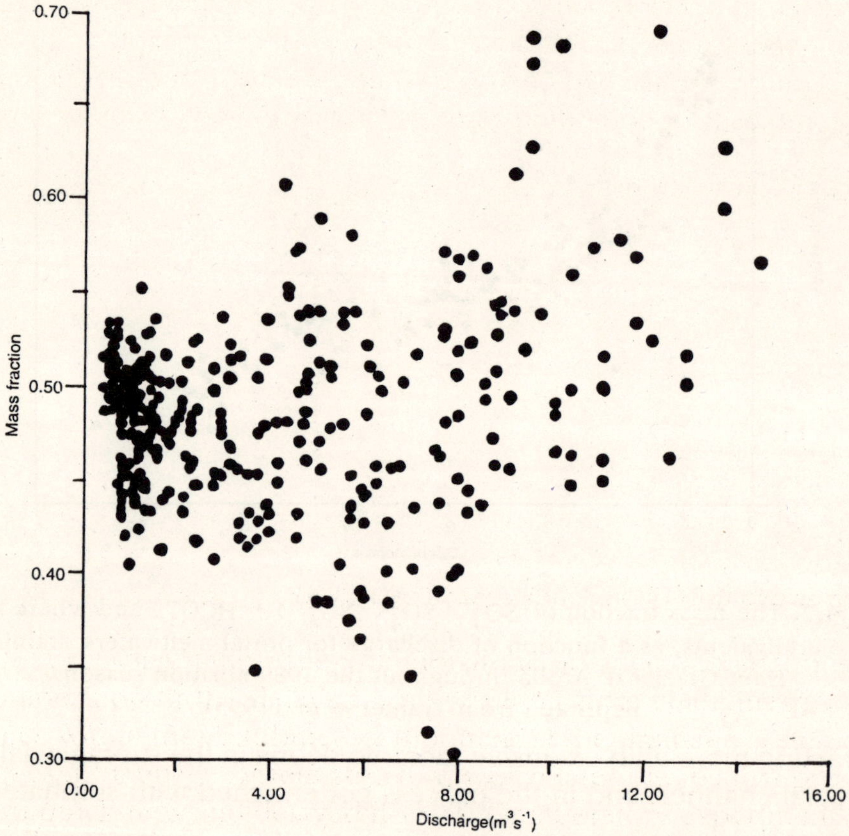

Fig. 4.8. The mass fraction of SO_4^{2-} (SO_4^{2-} / SO_4^{2-}) + HCO_3^-) and where units are equivalents, as a function of discharge for portal meltwaters draining Dokriani glacier, Garhwal Himalaya throughout the 1994 ablation season.

5

SEDIMENT TRANSFER PROCESSES IN GLACIERIZED BASIN

5.1 INTRODUCTION

The processes of glacial erosion and deposition occur when glacier flows over rock beds (Boulton, 1996). Erosion process entails four primary mechanisms: abrasion, quarrying, subglacial fluvial erosion, and dissolution (Drewry, 1986). The last two are dealt separately in this book. By the middle of 19th century Forbes (1842) and Tyndall (1864) recognized that rock clasts embedded in glacier ice abrade the underlying rock bed. Later Chamberlain (1878) established that icc behaves as a deforming fluid rather than a rigid body. These observations helped (Gilbert, 1903, 1906) relate icc flow at glacier beds to mechanism of rock fracture and erosion. Both abrasion and quarrying involve the fracture and dislodgement of rock from the bed. In the case of abrasion, fracture results from large stress differences localized beneath ice-entrained rock fragments in frictional contact with the bed. In the case of quarrying, such stresses are induced directly by ice, usually over larger areas of bed and for longer periods of time. Abrasion, therefore, produces fine debris, ranging from fine silt to coarse sand, whereas quarrying generally produces larger rock fragments. Dislodgement of quarried materials requires that frictional forces be overcome by the combined action of water and sliding ice.

Most quantitative models of glacial abrasion (Hallet, 1976; Riley, 1982; Shoemaker, 1998) consider the primary agents of abrasion to be clasts embedded in the basal ice of glaciers. Recently, Alley *et. al.* (1987a, b) have suggested that the subglacial deforming sediment layer may effectively abrade underlying rocks.

The quarrying (plucking) is defined as evacuation of loosened rock fragments. It is an important erosional mechanism based on diverse geological evidences; (1) rock fragments on glaciers and erratics in deglaciated areas; (2) widespread evidence for subglacial rock fracture on proglacial bed rock exposures; (3) occurrence of quarried lee surface on boulders that are embedded in lodgement tills (Sharp, 1982a, b).

Quarrying also plays an important role in the generation and maintenance of bed roughness, which also affects the sliding speed and stability of temperate glaciers (Sharp, 1982). Unlike abrasion, which tend to smooth the bed because it is more effective near the tops of the bed protuberances than in depressions, quarrying can roughen the bed, thereby permitting the bed micro-relief to persist inspite of extensive erosion. For instance, considerable bed roughness is evident on recently deglaciated bedrock in deeply incised valleys where erosion can amount to hundreds of meters of surface lowering (Hallet, *et al.*, 1996) also considers quarrying as a key erosive process because a sustained source of abrasive material active over large part of glacier bed is necessary for significant abrasion. Quarrying also affects the accumulation of subglacial sediments, which can influence sliding over bedrock (Shoemaker, 1986). Many recent field studies and models confirm that sediments, which can deform or facilitate sliding at the base of ice masses, may play an important role in the glacier dynamics (Boulton, 1987; Alley *et. al.*, 1986). Iverson (1991) computed stresses both in the ice and in the rock by using coupled finite-element models. He computed ice loads on a single bedrock ledge. Whereas, the model developed by Hallet (1996) computed stresses averaged over those parts of bed in contact with ice and also considered a series of ledges, instead of single ledge-cavity. The model can evaluate the rate at which microcracks can grow, which is in fact linked to the rate of quarrying.

5.2 CHARACTERISTICS OF THE SEDIMENT SYSTEM IN GLACIERIZED BASINS

The sediment production in the glacier system may be viewed as being dependent on the strength-stress relationship between interface materials and processes. There are primary production processes such as plucking, cavitation, gelifraction and hydrolysis and the secondary processes which rely on another process to achieve erosion such as abrasion, shearing and corrosion. The relative importance of the various processes in terms of their quantitative contribution to the overall sediment budget of the basin is not well understood (Fenn, 1987). Perhaps the sediments are produced differentially according to variations in the efficiency of processes like moving ice, water and debris as well as variations in the resistance of bedrock material.

5.2.1 Specification of the System

Glacier basin sediment transfer system is readily represented schematically (Figure 5.1). In essence, sediments may be derived from weathering processes and from glacial fluvial erosion processes from three main sources (subaerial, subglacial and proglacial) and may be routed by glacial, fluvial and mass movement transport processes along five main paths (subaerial, supraglacial, englacial, subglacial and proglacial). The various sediment production and transport processes may be represented, respectively, by the following expressions:

$$S_p = G_p + F_p + W_p \qquad (1)$$

and

$$S_t = G_t + F_t + M_t \qquad (2)$$

where the subscripts *p* and *t* refer, respectively, to sediment production processes and sediment transport processes, and where *S* denotes total sediment volumes and *G*, *F*, *W* and *M* denote, respectively, glacial, fluvial, weathering and mass. The presence of a mass of (usually) impenetrable overlying ice renders the subglacial environment 'uniquely inaccessible to direct or remote study' (Boulton, 1982). Information on the characteristics of basal ice, subglacial surfaces and interface processes has been collected from studies undertaken in natural subglacial

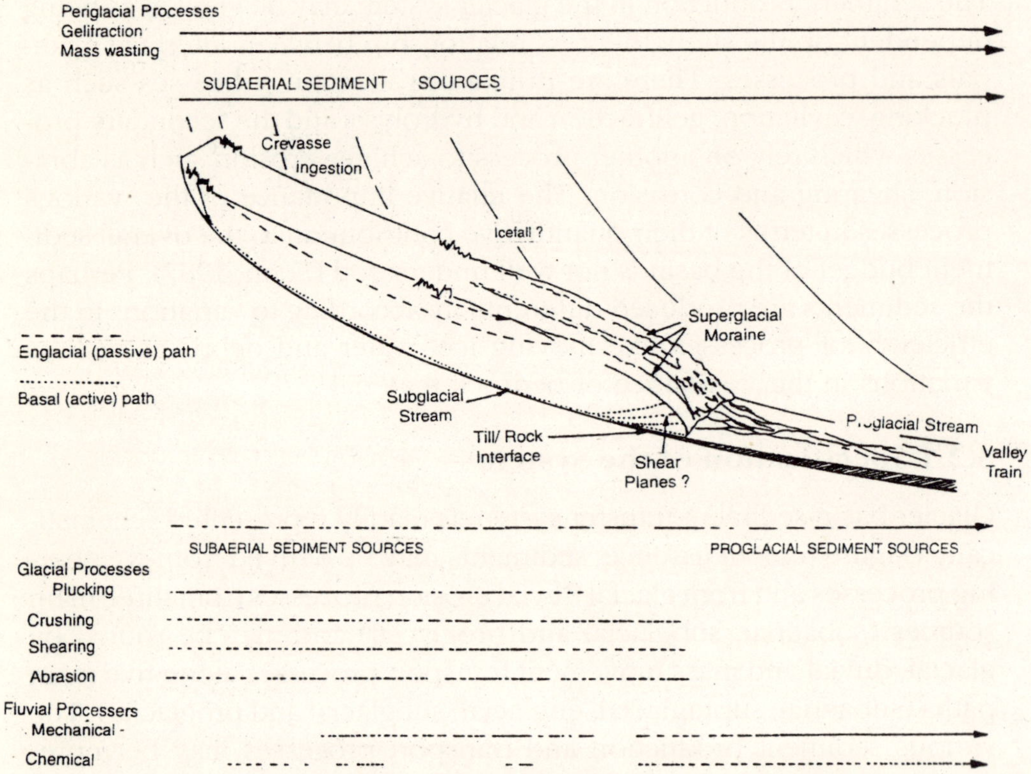

Fig. 5.1. A composite model of the sediment transfer system of alpine glacier basin. Reprinted from Fenn (1987)

cavities and in man-made tunnels and boreholes (Vivian, 1980). Observations from such sites are invaluable but, inevitably, insufficient: as they provide a view from a small window.

The term G_p includes abrasion, quarrying and fracture / crushing shearing of subglacial surfaces. F_p includes water-related mechanical and chemical erosion of subglacial and proglacial surfaces. W_p includes the various processes involved in the physical disintegration and chemical decomposition of rock materials.

The term G_t includes materials transported along the surface (via moraines), in the body (particularly via debris-charged 'septa' or debris layers) and at the base [in the 'sole' or basal ice layer] of the glacier. Following Boulton (1975, 1978), a distinction may be drawn between the 'passive' englacial pathway and the 'active' basal transport zone. The term F_t includes the transport of sediments as dissolved, wash, suspended and bed load in subglacial and proglacial sheet and channel flow. M_t includes mass movements of the fall, slide and flow type (rockfalls, debris slides, debris flows, slush avalanches and snow avalanches) on the rock faces and scree slopes flanking the basin both above and beyond the glacier.

Therefore, the glacial system is a complex erosion-transport-sedimentation system in which the various sediment production and routing processes operate interdependently rather than independently, differentially rather than uniformly and intermittently rather than continuously. Moreover, the system invariably incorporates lags and sediment stores: not all of the sediments produced during a given time span are necessarily removed concurrently, and not all of the materials removed from the basin in a given period are necessarily produced contemporaneously. Expressed as a simple balance equation taking storage into account, the transfers involved reduce to

$$S_s = G_p + F_p + W_p - G_t - F_t - M_t \quad (3)$$

where the term S_s describes sediments 'stored' in the basin (i.e. materials produced, but not removed, in past periods). The term in the above expression refers to sediment transported from the basin.

In most partially glacierized basins, however, the morphology of the catchment is such that sediments are exported as bed, suspended, wash and solute load in the proglacial streams. In balance terms, glacial and mass movement outputs may accordingly be treated as inputs to fluvial transport processes. Further, given the protracted transit times of the G_t and M_t pathways (whether involving slow, continuous transport or rapid, but episodic transport) the term S_s may be taken to include sediments 'stored' on subaerial rockfaces and scree slopes, on supraglacial surfaces and in proglacial valley trains. Equation (3) accordingly reduces at basin level to:

$$S_s = G_p + F_p + W_p - F_t \quad (4)$$

This indicates that changes in the basin sediment store are controlled by the balance between the volume of sediments produced by active glacial, fluvial and weathering processes and the volume of sediments removed by active fluvial processes. Restated, noting the $F_t = S_Q$, the total sediment discharge from the basin, Equation (4) becomes

$$F_t = S_Q = G_p + F_p + W_p + S_s \quad (5)$$

This indicates that the volume of material removed from glacier basins (as stream sediment load) is determined by the quantity of sediments delivered by active glacial, fluvial and weathering processes and by the volume of sediment released from or lost to sediment stores in the basin.

Fenn (1987) has proposed a model of the sediment transfer system as shown in Figure. 5.2. The model illustrates the sediment sources, and pathways.

5.2.2 Sediment Production

The efficiency of the erosive processes operating in glacial environments has long been a subject of debate. On one hand, there exists persuasive morphological and sedimentological testimony to the power of alpine glaciers, and evidence in the high specific sediment and solute load presently exported from highly debris-mantled glacier basins to indicate that present-day glaciers are at least acting as efficient debris transport systems if not as efficient debris production systems (Small, 1987). On the other hand, geomechanical evaluations of the relative strengths of ice and rock (McCall, 1960; Glen and Lewis, 1961; Carson 1971; Morland and Morris, 1977; Addison, 1981). Similar observations have been placed on the geomechanical effectiveness of gelifraction in field conditions (Ives, 1973). The situation, at first sight, bears some analogy to the aerodynamic objections to the flight of the bumble bee! The apparent dilemma is eased, however, when the problem is tackled from a whole-system perspective, admitting the interdependency of processes, differential erosive efficiency and spatial (and temporal) selectivity in sediment production.

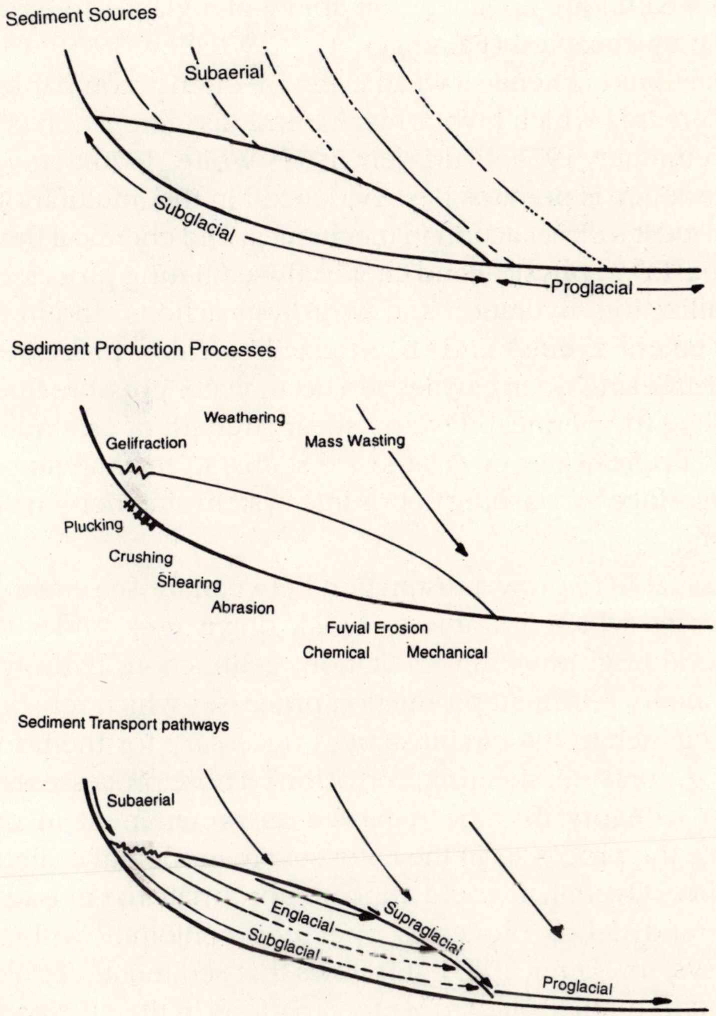

Fig. 5.2. Sediment sources, production processes and transport path. Reprinted from Fenn (1987)

5.2.3 Interdependency in Processes

Proof that processes operate in an interactive rather than in an isolated fashion is plentiful: (1) that gelifracted or plucked debris must be supplied to the basal ice transport zone before abrasion (and possibly crushing and shearing) may occur (Carson, 1971; Boulton, 1974); (2) that preglacial weathering enhances the ability of a glacier to erode its bed is increasingly accepted (Feininger, 1971; Whalley, 1979); (3) that the various reactions of chemical weathering are often accompanied by rock volume increases which favour mechanical disaggregation is also well known (Feininger, 1971; Brunsden, 1979; White, 1976). However, the interdependency is perhaps best evidenced in the multifarious role of water beyond it s direct action in mechanical and chemical fluvial in the erosion: viz. (a) in physical and chemical weathering processes (e.g. *in situ* crystallization, hydration and hydrolysis actions, and in triggering mass movement events) and (b) in glacial erosion processes (e.g. in promoting freezing-on in cavities subject to water pressure fluctuations; in controlling the permeability and shear strength of ground materials, whether via pore-water or ground-ice status; in maintaining an active ice-rock interface by washing debris into system channels; in promoting basal sliding.

It is possible to draw a distinction between those primary production processes which are able to break down rock under their own actions alone (e.g. plucking, cavitation, gelifraction, hydrolysis) and those secondary sediment production processes which rely on another process to generate the circumstances necessary for them to achieve erosion (e.g. abrasion, shearing, corrosion). Those processes falling into the former category may be regarded as pre-eminent in qualitative sense, since the processes in the latter set are inoperative/ineffective in their absence. Overall, it would appear appropriate to envisage moving ice, water and debris processes operating conjointly within a basin erosional system (Fenn, 1987). It follows that sediments are likely to be produced differentially according to variations in the efficiency of processes as well as variations in the resistance of bed materials.

5.2.4 Subglacial Sediment Production

At a general level, the efficiency of sediment production may be viewed as being dependent on the strength-stress relationship between interface materials and processes. It has long been accepted that glacier may efficiently exploit inherent rock mass instability, and the erosion is, in consequence, likely to be spatially uneven with sediments being produced preferentially from 'soft' sources [i.e. discontinuous and weathered rock mass zones (Carson, 1971; Addison, 1981)]. It is also accepted that there are likely to be particular points or zones underneath a glacier at which the forces exerted on the bed are greater than they are elsewhere. Recent research at subglacial sites has shown just how effective a glacier may be when circumstances combine favourably. The findings of Robin (1976) and Hagen *et al.*(1983) are particularly interesting in this respect. Robin (1976) showed that basal shear stresses may increase by around 10 bar in cold patches within temperate ice zones. Hegen *et al.*(1983) observed substantial variations in basal ice pressure under Bondhusbreen, Norway: (a) that unsteadiness during basal sliding produced average basal pressures 3-4 bar higher than hydrostatic pressure; (b) that large variations in basal pressure accompanied the growth and closure of subglacial cavities produced by water pressure fluctuations; (c) that huge rises in basal pressures were produced by boulders passing through the basal transport zone (pressure rises of up to 90 bar were recorded); (d) that basal pressures varied spatially with differences of up to 30 bar having been measured. It would appear, therefore, that effective ice pressures may be considerably enhanced by dynamic effects associated with glacier sliding, entrained debris and thermal behaviour to levels well beyond the theoretical maxima assumed in steady-state geomechanical tests. It follows that the forces available for erosion may, at least locally, be much higher than would be expected from traditional temperate ice theory.

It seems clear that the absolute and relative efficiencies of subglacial erosive processes vary according to spatially variable bed conditions (structure, lithology and morphology) and spatially and temporally variable 'process' conditions (flow type, flow rate, basal ice temperature, subglacial drainage). Zonal variability in erosion has also been modelled, e.g. Boulton (1974) attached erosional and depositional attributes to zones defined according to thermal regime. Clayton and Moran (1974) advanced a glacier form-process model based on thermal, groundwater and flow regimes; while Andrew's (1972) concept of glacier power embodies a flow-based distinction in erosive potential, with effective power increasing as a proportion of total power (a maximum condition) as basal sliding increases as a proportion of total glacier flow.

5.2.5 Sediment budget and balance

It is possible in generalized, qualitative terms, to identify the major controls on the quantity of sediment exported from a glacier basin. Equation (5), for example, may be linked to the water balance equation of drainage basin. However, it is not so easy to quantify process-specific, source-specific or time-specific contributions to sediment outputs (there are parallels here with the difficulties involved in deconvoluting riverflow hydrographs). A number of studies have, however, provided indications of the importance of contributions from particular processes and/or origins. Examples relating to specific issues are as follows:

(a) On the relative importance of particular mechanisms. Boulton (1974) and Boulton *et al.*, (1979 a,b) argued that plucking is likely to be quantitatively more important than abrasion in circumstances where supraglacial debris inputs are limited as in French and Swiss Alps; the volume of sediments produced by plucking is unlikely, however, to exceed that resulting from abrasion in valley glacier situations, where valley walls may be expected to deliver substantial quantities of debris on to and into the glacier as in Karakoram and Himalaya.

(b) Recent studies of the cations load of proglacial streams indicate that current rates of cationic denudation in glacial catchments are noticeably higher than the continental mean rate quoted for, Gornergletscher (Collins, 1983), Berendon (Eyles *et al.*, 1982) and South Cascade (Reynolds and Johnson, 1972) glaciers respectively. Whether the action of flowing water directly produced the breakdown of rock (i.e. erosion *senu stricto*) or simply removes material produced by other (glacial, cryergic or, weathering) processes, its influence is evidently manifest. Hagen *et al.*, (1983), for example, noted that around 90% of the material passing through a cross-section monitored under Bonhusbreen Norway, was carried in meltwaters rather than by the glacier itself.

(c) Sediment delivery in glaciated basins seems to be highly variable according to geological, glaciological, geomorphological and hydrological circumstances. Mills (1979), for example, reported that most of the Nisqually rivers suspended sediment load appears to be entrained by the time it emerges from the Nisqually glacier, with over two thirds of the total sediment load being derived subglacially. The Storbeen catchment in Norway, in contrast, is reportedly characterized by a discontinuous process of sediment transfer from subglacial sources to proglacial sediment stores during melt-dominated periods, and by evacuation of sediments from the proglacial zone during rainfall controlled high flow events (Richards, 1984). Perhaps the most comprehensive sediment source area results produced to date are those of Hammer and Smith (1983). Their studies in the Hilda glacier basin, in Canada, enabled them to estimate the proportion of bedload and suspended load delivered to the stream from supraglacial, subglacial and proglacial sources: 46 % of the bedload was deemed to be of supraglacial origin, 27 % of subglacial origin and 27 % of proglacial origin; only 6% of the suspended load was of supraglacial origin, but with subglacial and proglacial sources each yielding 47 % of the total load.

(d) On past and present production contributions to present sediment output. A number of studies have derived estimates of rates of denudation in glacial catchments on the basis of sediment output in proglacial streams and/or from marginal moraines (Ostrem, 1975; Small *et al.*, 1984). Exactly what proportion of the material presently transported out of a glacier basin in proglacial streams is the result of currently active erosion processes is a matter of some uncertainty. Present sediment outputs strictly represent present sediment removal and not necessarily present sediment production. Views on the relationship between present sediment inputs and outputs vary widely. Metcalf (1979), for example, took the annual suspended sediment load to be an order of magnitude approximation to the annual mass of subglacial abrasion as a general premise n a sediment budget study of the Nisqually glacier basin, USA. In discussion of Collins work Boulton (1979 c) commented that long-term net increases or net losses of sediments in subglacial stores are unlikely to be large in most cases. Conversely, however, Church (1974) reported that the sediment yield in glacier basins in Baffin bears to present primary rates of sediment production, a condition which has been taken by Church and Ryder (1972) to characterize 'paraglacial' environments. While sediment budget and balance characteristics undoubtedly differ from basin to basin, it is probably safe to assume the existence of some storage of sediments in all basins: the magnitude-duration characteristics of debris stores may vary from basin to basin, but there is always likely to be an element of storage present (Fenn, 1987).

5.3 RATES OF EROSION

The most reliable estimate of erosion rates comes from measurement of sediment discharge in meltwaters streams. A major assumption associated with these measurements is that over the period of measurement, water has access to most of the glacier bed so that there is negligible storage of debris. Furthermore, it is assumed that most quarried debris is sufficiently comminuted so that it can be fluvialy transported. Very few data are available with which to assess the relative importance of abrasion and quarrying, but quarrying is probably dominant (Drewry, 1986).

However, monitoring of drainage basins sediment yield provides only an integrated view of how sediment respond to average conditions. Measurement of individual process and the construction of a sediment budget can provide additional information about the sediment system (Walling, 1983).

A total of 1.79 Gt year^{-1} of suspended sediment (about 9 % of the total annual load carried from the continents to the oceans worldwide) is transported in three major river systems namely: Brahamputra, Ganga and Indus in a combined runoff of 1.19×10^3 km^3 (Meybeck, 1976). About 75% of the runoff in these three major rivers occurs between June and September, by the monsoonal precipitation, snow and glacier icemelt (Collins and Hasnain, 1995). The effect of the monsoonal precipitation provides much accumulation on the Himalayan glaciers in summer when strong ablation occurs under strong solar radiation. Sometimes, summer snowfall changes to rain under warm air temperature condition, which accelerates new snow cover melts. Consequently, summer air temperature in the southern Himalaya is an important factor which controls ablation through albedo variation at the glacier surface. These glacier regimes have been called the summer-accumulation type. Their characteristics have been described by Ageta and Higuchi (1984). Relatively young age of Himalayan-Karakoram mountains with large and active glaciers, high seismicity, steep valleys with frequent avalanching and intense monsoonal rainfall are contributing to the erodability of the slopes (Hasnain and Chauhan, 1993). Himalayan glaciers produce very rich debris and are relatively thick and have large lateral and end moraines compared with other glaciers in the world. Hence, moraine-dammed lakes are commonly formed by the retreating glaciers. They are unstable and a slight disturbance by seismicity or cloud burst can easily alter the balance of the structure and an abrupt release of stored water with debris may result (Yamada, 1993). High speed, erosive glaciers driven by high accumulation rates, steep topographic gradients, high rates of water loss in their ablation areas, and yielding high concentration of sediment in their meltwaters. Such glaciers are thought to be active agents and linkages in regional denudation system (Gardner, 1986).

5.3.1 Erosion Rates in Himalaya

The climatic conditions have a substantial influence on sediment evacuation (Collins and Hasnain, 1995). In Table 5.1 sediment transport data of six months ablation period of Dokriani glacier, Garhwal Himalaya are shown. A total of 15×10^4 tonnes suspended-sediment was transported. The months of July and August (monsoonal) contribute 64 % of the total discharge, 70 % of the total suspended transport and experience 73% of rainfall (934 mm). During June, 26% of the total suspended-sediment was transported, though the discharge was less than that in August. This is probably due to the evacuation of preceding summer/winter sediments from the subglacial region.

5.4 INTER–RELATIONSHIP BETWEEN FLUVIAL SEDIMENT TRANSFERS

Warburton (1990 b) has proposed a model showing the main elements and their inter-relationship with transport via mainstream channel. The framework he proposed refers to processes occurring during the ablation period. The schematic diagram of the model is shown in Figure 5.3.

Table 5.1

Month	Total discharge (10^4m)	% Discharge	Sediment flux (10^4 tonne)	% sediment flux	Rainfall (mm)	% Rainfall
May	195.5	3.0	.04	.028	4.0	
June	868.0	14.0	4.0	26.0	15.0	9.0
July	2187.0	35.0	8.0	52.0	500.0	39.0
August	1785.0	29.0	3.0	18.0	434.0	34.0
September	910.0	14.0	0.52	3.5	189.0	15.0
October	292.0	5.0	.02	0.19	26.0	2.0
Total	6238.0		15.0		1270.0	

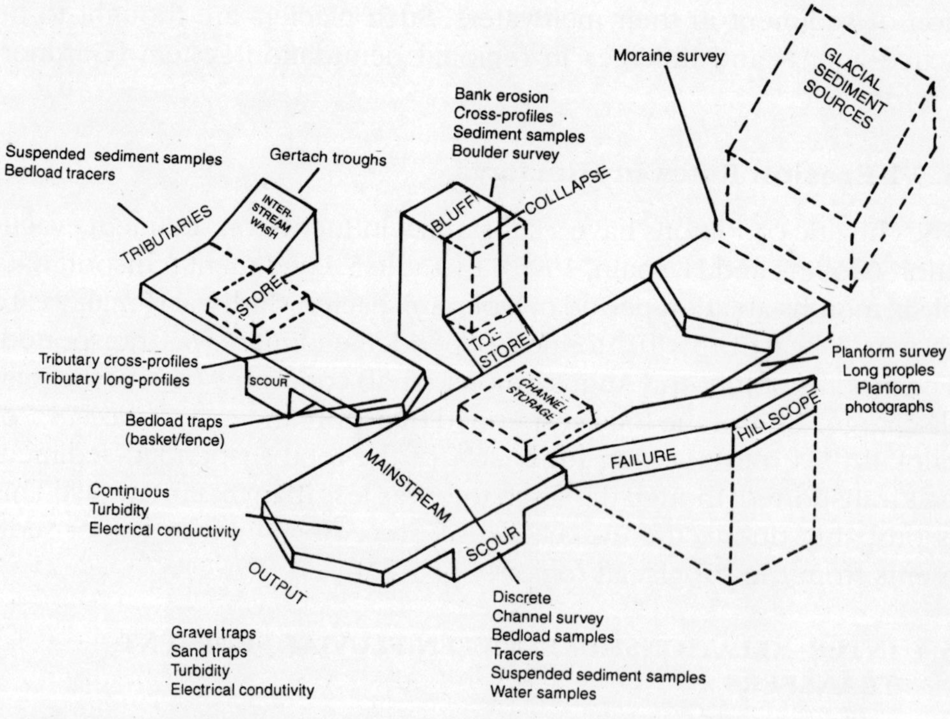

Fig. 5.3. Inter-relationship between fluvial sediment transfers and field measurements. Reprinted from warburton (1990)

5.3.1 Sediment Balance Equation for a Proglacial Stream

The basic equation for the proglacial budget can be written as follows:

$$Y = S_L + T_R + M + G_L + V_S$$

where Y is sediment yield; S_L are the direct hill-slope inputs; T_R are tributary stream channel inputs from moraine deposition; G_L is the contribution from glacial stream sources; and V_S is the change in valley sediment storage. The terms Y, S_L, T_R M and V_S can be measured in the field. However, the only term to be estimated is G_L which can be calculated as the difference of other terms. Based on this equation total sediment output from the basin can be derived from proglacial sediment sources. This provides solution to the problem of measuring sediment yield directly at the glacier snout where the meltwater stream emerges from under the glacier.

6

SUSPENDED-SEDIMENT TRANSPORT IN GLACIAL MELTWATERS

6.1 INTRODUCTION

A number of outstanding issues in the earth sciences can be addressed by studying sediment production from glaciers and ice sheets. These issues include the role of sediment in glacier dynamics, the characteristic of subglacial hydraulic systems and the processes of erosion, sediment transport and deposition (Merrand and Hallet, 1996). How erosive are glaciers ? What is their role in mass transfer in glacierized regions of the Himalaya-Karakoram mountain system ? These are some of the fundamental questions that need to be addressed in understanding the regional denudation processes in the Himalaya.

Meltwaters draining from the temperate glaciers transport large quantities of suspended-sediments. The suspended-sediment concentration in meltwater is highly variable at both short-term and seasonal time-scales. This variability has been related to availability of sediment supplied to the subglacial channel system (Ostrem, 1975), changes in the area that water is washing over during a flood event (Liestol, 1967) and exhaustion effects (Gurnell, 1987, Gurnell *et al.*, 1994). Collins (1989) interprets the seasonal pattern of variation in sediment flux of meltwater from Gornera glacier, Switzerland as a seasonal development of the subglacial drainage system. Throughout the season, areas of previously hydraulically isolated subsole are integrated with flow, releasing quantities of sediment from basal storage. Thus, the large amount of debris transported suggests that meltwaters flow at the subsoles of glacier for much of the distance between points of entry on a glacier surface and the ice margin. At subsole, sediment produced by glacial erosion processes is picked up by flowing meltwater. Entertainment of products of abrasion is unlikely to be uniformly distributed across the subsole as meltwater usually emerges from a glacier in distinct filaments of flow which probably cover only a small proportion of the basal area. Also, not all filaments of flow will have sufficient velocity to entrain all sizes of particles, so that where faster flowing meltwater is absent, sediment will accumulate at the subsole.

Transport of debris from glacier subsoles by meltwater is irregularly distributed in time. In winter, overall discharge and velocities in individual branches of the drainage network are too low for entertainment of even fine sediments. As flow rises, with increasing air temperature in spring and early summer, disproportionately large quantities of suspended sediment are transported by comparison with those moved by higher flows later in the season. Major spatial instabilities of the basal drainage network punctuate the seasonal pattern of sediment flux particularly in the early part of the ablation period.

Concentration of suspended sediment (commonly expressed as parts per million, milligrams per liter, or grams per cubic meter of water) and the amount of transport material show great variations with time in meltwater stream; perhaps these variations are greater and occur more rapidly than in any other type of stream. Diurnal variations of sediment transport from a Himalayan glacier during a typical monsoonal day in September 1992 is shown in Figure 6.1. The peak in sediment concentration occurs earlier than the water discharge peak. This is a very common phenomenon found in all meltwater streams, for samples taken close to the glacier snouts.

6.2 SUSPENDED-SEDIMENT CONCENTRATION-DISCHARGE RELATIONSHIPS

In order to assess the nature of sediment supply beneath the glacier and the interaction of water flow with sediment at the bed, the relationship

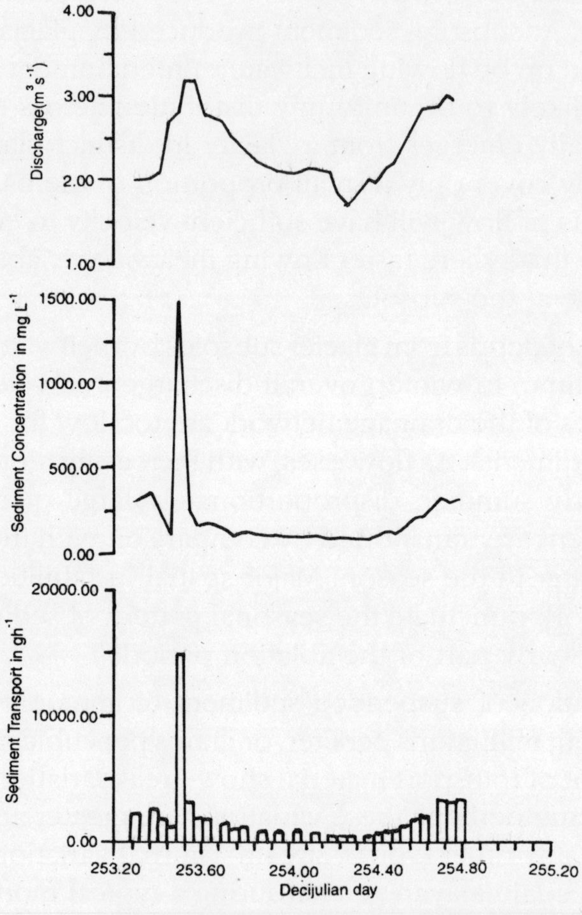

Fig. 6.1 Diurnal sediment transport during a typical monsoonal day in September 1992. Reprinted from Hasnain (1996)

between suspended-sediment concentration and discharge should be studied in detail. The suspended-sediment concentration-discharge relationship is of the form:

$$S_s = aQ^b$$

where

S_s = Suspended sediment concentration
Q = discharge

a, b = best fit estimated parameters, using r^2 as an index of goodness of fit of the model.

The model was fitted to the data collected at Dokriani glacier, Garhwal Himalaya during August-September 1992 (Hasnain, 1996), as shown in Table 6.1. The goodness of fit index (r^2) varies both in the rising and falling limb of the diurnal hydrograph indicating that the relationship between sediment concentration and discharge is non-stationary. The effect of hysteresis dynamics in the relationship between sediment concentration and discharge is shown by the concentration loop for the 9-10 September 1992 (24 h) continuously hourly recording on Dokriani glacier, as shown in Figure 6.2. The relationship between sediment concentration and discharge in the Gornera glacier, Swiss Alps is illustrated by rating loops for 18-19th August 1975, shown in Figure 6.3. The open-ended loop indicates that meltwater takes various hydrological pathways and conditions changing from one day to another. Clockwise hysteresis loops result from greater concentration of suspended sediments on the rising limbs of the diurnal hydrographs than at equivalent discharge during the falling stage, resulting from the flushing out of sediment collected at margins and beds of subglacial streams during low flows. A distinct loop occurred each day during periods of sustained ablation, often with figure-of-eight sub-loops and involution's, reflecting irregularity in sediment availability.

Table 6.1 Suspended sediment concentration-discharge relationship, determined for rising and falling limbs of daily hydrograph during the 1992 ablation season

Date	Hydrograph limb	a	b	r^2
28 Auguust 1992	Rising	–0.78	0.29	0.6
29 August 1992	Rising	–0.91	0.36	0.91
30 August 1992	Falling	–0.26	0.16	0.87
31 August 1992	Falling	–0.09	0.10	0.83
03 September 1992	Rising	–4.41	1.46	0.61
05 September 1992	Falling	–0.22	0.15	0.85
06 September 1992	Rising	–0.19	0.14	0.91
09 September 1992	Falling	–0.48	0.28	0.91
10 September 1992	Falling	–0.03	0.06	0.79
11 September 1992	Falling	0.29	0.19	0.84

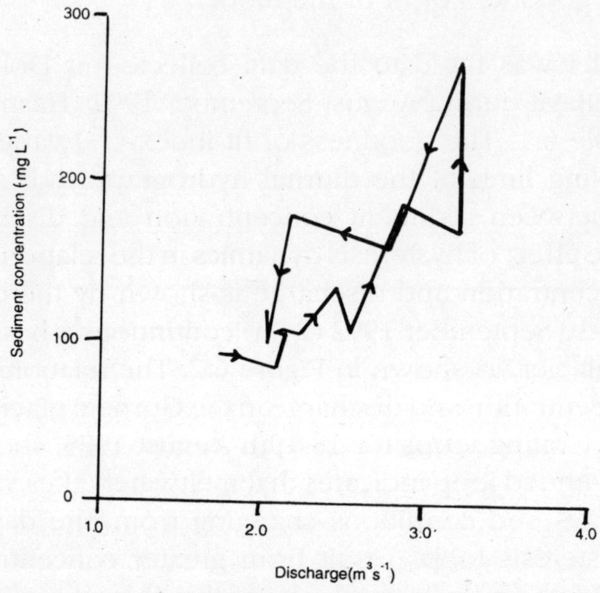

Fig. 6.2. Hysteresis rating loop of suspended sediment concentration with discharge in the Dokraini glacier for 9-10 September 1992. Reprinted from Hasnain (1996)

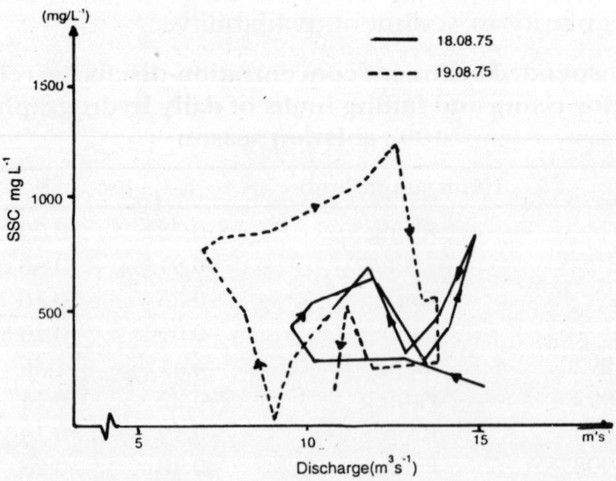

Fig. 6.3. Hysteresis loops for suspended sediment concentration with discharge in the Gornera glacier for 18 and 19 August 1975. Reprinted from Collins (1979a)

6.3 SUBGLACIAL SEDIMENT SUPPLY AND DELIVERY TO MELTWATERS

Sediment will be produced by glacial erosion processes throughout the year, whereas meltwater flow with sufficient velocity and turbulence maintain traction and suspension during the ablation season. Under such conditions it is very unlikely that the sediment would continue to accumulate in particular areas of the bed. Consequently, debris will be removed constantly by the migration of subglacial channels over wide areas of glacial beds. Sediment will be released to meltwaters in the basal hydrological system by melting of debris-rich basal ice from the glacier sole, and by deformation of basal sediment into passageways through glacier sliding. Migration of channels with sliding ice, and movements of channels relative to surrounding ice by melting of channel margins, may allow meltwaters to impinge on patches of basal sediment.

Sediment transport to the glacier portal in a period of time will be supply related. It depends on how much area of unworked glacier sole becomes integrated with flow and how much sediment is held there. The other controlling factors are the thickness and debris content of basal layer, nature of bedrock, presence or absence of unconsolidated, deforming sub-sole. Areal integration of sole with flowing meltwater will depend on the type of basal conduits and on how the system is evolving in response to changes in water pressure arising from seasonal and diurnal variations of surface-water input. According to Collins (1989) the bulk of sediment delivery probably results from glacier sliding by bringing basal ice and deforming basal sediments from up-glacier into contact with flowing meltwaters or relocating conduits incised upwards into ice on the areas, basal sediment down-glacier. These interactions will be more effective the more transverse the orientation of conduit system is to the direction of sliding. A dense network of small conduits arranged at an angle to the direction of glacier flow would be expected to gather more sediments than a few large conduits with alignment parallel with flow. Development of conduit system in this way would explain the decline in sediment flux with progression of ablation season. Variations in sliding velocity, together with changes in conduit system, may alter the area of debris contact with flowing water, and will influence sediment load.

6.3.1 Subglacial Hydrological Events

Relationship between discharge and suspended sediment load in total meltwaters becomes very complex by the occurrence of subglacial hydrological events. Level of discharge which exceed previous flows of the season do not always have an impact on sediment flux. There is no relationship between the quantity of sediment evacuated during an event and discharge magnitude. The impact of subglacial hydrological event is related to timing of occurrence and position in the seasonal series, and also to the area of sole involved.

The sudden large-magnitude variation of sediment concentration suggests subglacial instability beneath a glacier ice. Random injection of sediment impulse may result from migration of basal streams during reorganization of subglacial drainage channels. Sediment impulses were often associated with the presence of large quantities of cobble-sized ice particles floating in meltwaters as could be noticed in Gangotri glacier, Garhwal Himalaya, India during June and July, indicating the reorganization of subglacial channels on a large scale causing the fracture of large columns of basal ice. During migration, channels impinge on unworked pockets of sediment stored at the ice-bedrock interface, providing ready but short-lived supplies of sediment load. Slumping from unstable moraine conduit margins may also produce sudden increase in sediment concentration followed by rapid exhaustion.

6.4 SEDIMENT YIELDS OF GLACIERS

Sediment yields have been shown to be different for various types of glaciers. With some exceptions, large valley glaciers are characterized by large yields, and small glaciers by low yields. Glaciers in different regions having different thermal regimes and on various bedrock types are also characterized by different yields (Hallet *et al.*, 1996). Sediment yields are obtained by monitoring of water discharge and sediment transport in meltwater streams close to glacier outlets. Sediment samples may preferably be collected with automatic samplers programmed for four samples a day.

A model was proposed by Bogen (1996) where the sediments are introduced to the subglacial waterways by melting of debris-rich ice in the glacier sole as shown in Figure 6.4. The rate of deformation of conduits depends on the difference between the ice overburden pres-

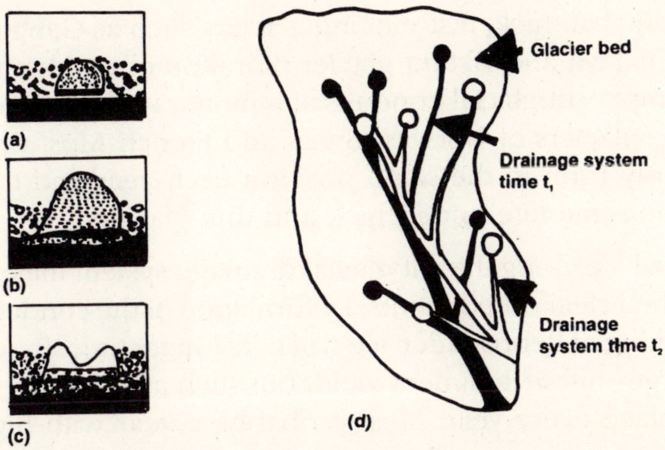

Fig. 6.4. Schematic diagram indicating the way that a cavity or an ice tunnel melts out sediments incorporated in the ice, (a, b) water pressure is rising and the subglacial cavity is expanding. (c) water pressure is falling and the tunnel or cavity is deformed. (d) changes in the position of the drainage system may increase availability of sediments. Reprinted from Bogen (1996).

sure and the internal water pressure. Due to variation in glacier melting and precipitation, water discharge in the subglacial conduits is subject to changes throughout the season. In periods of low water discharge, the movement of the glacier and the plastic deformation of the ice will deform the conduits. A subsequent expansion by an increase in water pressure will melt more ice and add more sediments to the subglacial system. In a system of equilibrium conduits, the dominant sediment sources are the beds of the conduits and possible subglacial till. Such sources should be expected to give good relation between sediment load and discharge. The pattern of sediment transport in the meltwater rivers of the glacier is availability controlled. That is, the rate of increase in discharge does not always give the same rate of increase in sediment load, depending on whether sediments are available for transport. This pattern, along with low erosion rate, strongly suggests that subglacial till is not the dominant source. Most probably, the glacier soles are resting on bedrock on which till is only present in patches. In such systems, fluctuations in water pressure and glacier flow rate are as important as total runoff in controlling the acquistition of sediment.

It is likely that thick, fast-moving glaciers such as Gangotri glacier, Garhwal Himalaya and Batura glacier, Karakoram mountains will deform their empty subglacial conduits at a more rapid rate than the thin, slow moving glaciers of Norway, Swiss and French Alps. Channels do not necessarily form in the same position each year and they do not deform at the same rate under thick and thin glaciers.

The basal ice along the subglacial drainage system may after some time be washed clear of sediments. Deformation of the conduits to allow for more melting of debris-laden ice will take longer time. In some years, there is a large shift in sediment yield, but such a shift does not necessarily take place every year. Most probably, a year with high yield is preceded by winter season ice deformation causing a major change in the position of conduits or linked-cavity systems to areas of ice richer in sediment. In some years the sediment yields remain low, presumably because no change in position took place. Therefore, the melt-out processes that remove debris from the sole operate on a much shorter time-scale than glacier erosion. In order to use records of sediment yields in meltwater to estimate erosion rates, several years of data should be used to even out variability originating from melt-out processes.

6.5 MODEL OF MELTWATER-SEDIMENT INTERACTION

6.5.1 Subglacial Entrainment of Sediment by Meltwater

Collins (1995) has proposed a conceptually based model of meltwater-sediment interaction at the base of the Swiss glacier. According to him high magnitude sediment flux events occur each time discharge rises to exceed previous levels, and filaments of flow beneath the ice successively expand the area of subsole covered by water to incorporate progressively parts of the subsole which have remained hydraulically isolated for sufficiently long time to allow large quantities of the sediment produced by glacier abrasion to accumulate. The events represent stages in the seasonal evaluation of the basal drainage system. The drainage network expands spatially in order to accommodate increasing flows of meltwater through individual reaches and to drain an expanding suface ice area as the transient snowline migrates in response to depletion of the winter snow cover. As discharge increases, filaments of flow widens and expand from the initial wetted area to integrate areas marginal to the channels. When flow declines with the return of cooler hydrometerological conditions, the wetted area contracts and filaments of flow presumably return to the same position on the bed that were occupied previously at that level of discharge. Removal of fine sediments from areas of subsole margins to a filament of flow will result from growth of the cross-sectional area. Evacuation of sediment at the bed and release of any debris in the basal ice layer by melting of the ice forming the roof over a channel are formed by the width of a filament of flow expanding at greter rate than the depth. For example in an elliptical cross section area as shown in Figure 6.5a, b, c or where channels widen by exploiting the sediment layer in preference to melting ice. Some increases in discharge, however, do not result in sediment flux events. In such events, the drainage net is not displaced from within the boundaries of the recently-swept area of subsole. Conversely, some areas of hydraulically-isolated subsole charged with sediment evidently become integrated with flow without an increase in discharge (Collins, 1995).

The measured flux of suspended sediment at times before, between and after the events requires continuous sediment delivery to basal meltwaters, although the actual quantity supplied declines through the season. Some sediment will be acquired from channel margins as diurnal variation in discharge changes the wetted area. Movement of filaments of flow with respect to stored sediment is brought about both by sliding of the channels across the subsole with the ice mass as a whole and by shifting of the channels with respect to the glacier through melting of tunnel ice-walls by flowing meltwater. Accumulated wet deforming basal sediment will also be squeezed into hydrological pathways at channel margins.

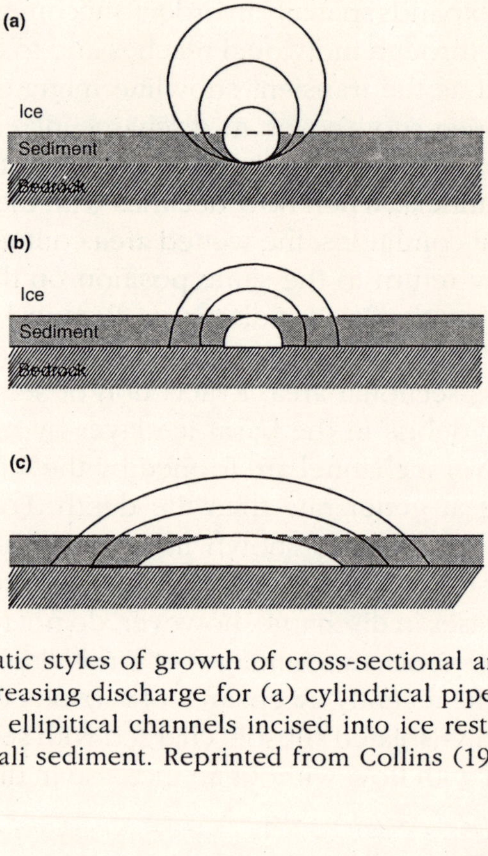

Fig. 6.5. Schematic styles of growth of cross-sectional area of subglacial pathways with increasing discharge for (a) cylindrical pipes, (b) semicircular channels and (c) ellipitical channels incised into ice resting on a layer of basali sediment. Reprinted from Collins (1995)

6.5.2 Model

The model consists of a grid-square lattice which represents the subsole of a rectangular valley glacier (Figure 6.6). The lattice is tilted at a low angle down glacier. In winter, the small water flux is restricted to flow along the long axis of the grid. An abrasion rate (producing A g day^{-1} $area^{-1}$ of uniform silt size) is applied uniformly in space and time to all cells in the lattice not occupied by meltwater in each one day time increment. Charging of a cell starts after the last inundation by meltwater in summer and continues throughout winter so that by the onset of melt in the following spring, substantial quantities of sediment will have accumulated across the entire subsole.

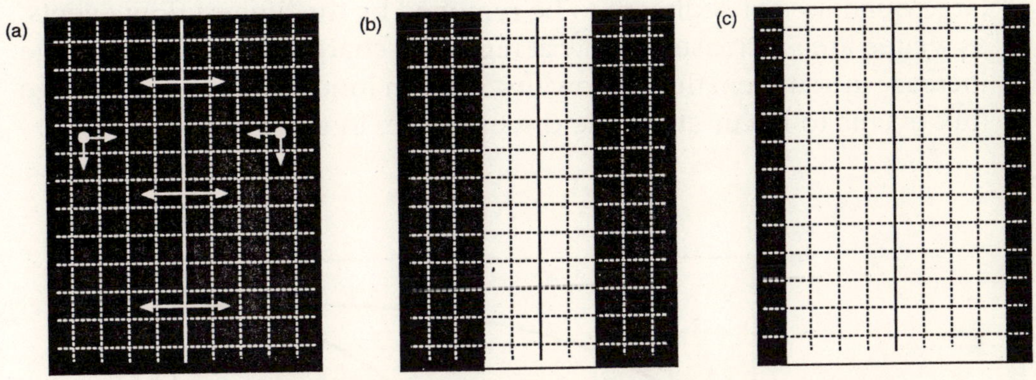

Fig. 6.6. Plan view of grid-square lattice representing the subsole of a valley glacier showing the direction in which the total wetted area expands equally on both sides. Reprinted from Collins (1995)

Increasing total wetted area with discharge in spring and summer is represented by the margins of the zone accomodating flow spreading out equally on both sides of and in parallel, with the long axis centerline. All the sediment accumulated in a cell is entrained when meltwater first occupies the cell, so that the wetted area becomes sediment free. Each successive expansion in wetted area produces a sediment flux event, the size of which is determined by the intergration of the quantity of sediment stored over the newly-wetted area. Where previous discharge levels arc exceeded by a small margin only, in mid-summer, and the newly-wetted areas are probably small, sizable sediment flux events may still result because the extended hydraulically isolated period will have allowed considerable sediment accumulation. Sediment will therefore continue to accumulate throughout summer in unwetted cells at the outer edges of the lattice. Cells from which winter sediment accumulation has already been flushed will be recharged during periods in summer when discharge remains low. Such small summer recharge quantities will be removed on the next sufficient increase in wetted area.

The relationship between the number of cells wetted and discharge has to be prescribed. Some possible rating curves in which total wetted area increases to less than 100% of the total number of squares at the maximum observed discharge are shown in Figure 6.7. Curve iii allows large quantities of sediment to be removed by the highest flow events, as wetted area increases rapidly at higher discharges. Curve ii gives little increase in sediment flux as flow first rises, in June, but then allows large flux events to occur at middle discharge i.e. in early July.

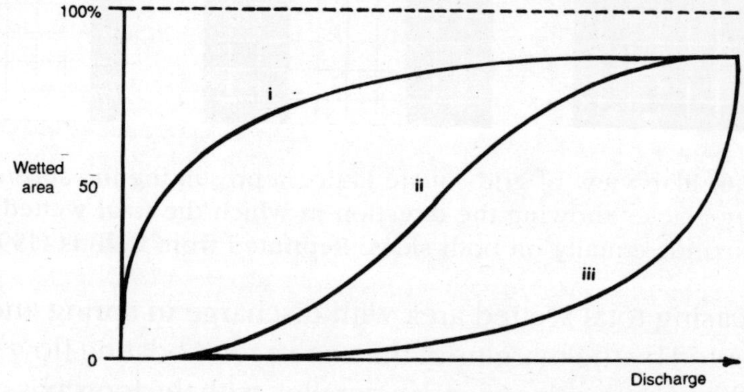

Fig. 6.7. Possible rating relationships between discharge and total wetted area of the subsole. Reprinted from Collins (1995)

6.5.3 Model Calibration

The model depends on the use of a suitable abrasion rate and on the hydraulic geometry which relates the overall wetted area of the subsole to discharge. An abrasion rate can be obtained by averaging total annual sediment flux from a glacier over several years and distributing the average flux uniformly over the area of subsole wetted at the maximum discharge. For calibration purpose it may be assumed that the sediment flux is derived from the entire subsole area. This will give a time-invariant daily rate of sediment production. Actually, the average abrasion rate must be higher, since, during the ablation season, glacial erosion will not be occurring over hydraulically active areas of the subsole, and not all the subsole area will be flushed by meltwater at maximum discharge.

The model confirms that both length of time since the previous occurrence of a particular discharge level and difference in magnitude of flow above previous high level in the season influence the amount of sediment transport during flux events. There remains considerable potential for the use of this simple conceptually-based model with calibration, in investigations of sediment-discharge relationships in meltwater draining from glaciers and of the impact of meltwaters on sediment accumulation beneath glaciers.

6.6 BASIN SCALE AND TEMPORAL VARIATIONS IN PARAGLACIAL SEDIMENTATION

In any argument suggesting that the lack of a present-day relationship between glacierization and sediment yields implies that rates of glacial and nonglacial sediment production are currently not equal, the most important implicit assumption is that there is no change in sediment storage. If this assumption is not true, then changes in sediment storage need to be estimated through sediment budgeting without which any conclusions about relative process rates from differences in short-term sediment yields cannot be drawn. To explain this in more detail we need to review some past and current ideas concerning paraglacial sedimentation and the relationship between sediment yield and basin area.

There are considerable empirical data available indicating that specific sediment yield decreases with basin area (Jansson, 1988). However, a number of recent studies have demonstrated that, contrary to this,specific sediment yield may in fact increase downstream, due to remobilization of sediment stored along a river (Ferguson, 1984; Warburton, 1990b). Church and Slaymaker (1989) and Owens and Slaymaker (1992) demonstrated that for British Columbian rivers specific sediment yield increases at all spatial scales up to 3 x 10^4 km^2. For undisturbed basins between 40 and 10^4 km^2 specific sediment yield is approximately proportional to $(area)^{0.6}$, thus implying increasing sediment yield downstream. It is also clear from the British Columbian data that those basins containing contemporary glaciers have the highest specific sediment yields (over basin scales of 1 to 50,000 km^2), although there is considerable variation about the trend, and variable surficial and bedrock geology must be factors contributing to this. This general conclusion is also supported by Walling and Kleo (1979) and Jansson, in a global survey of sediment yield from 1358 drainage basins with area ranging from 350 to 100,000 km^2, found that within a particular climate class where glaciers are active, sediment yields tend to be higher (although the comparability of the drainage basins is not known).

The reason for the increase in specific sediment yield with basin area in the British Columbian data appears to be that the major valleys are still responding to the last glaciation, and are actively eroding sediment delivered to the valleys by glacial events. Along similar lines, Ferguson (1984) observed that "most of the load of the Hunza River is picked up from valley floors even if it is ultimately of glacial orgin and transported by glacial meltwater". Church and Slaymaker (1989) present a modified version of the paraglacial sedimentation cycle (as initially defined by Church and Ryder, 1972) to accommodate a delay in response which increases with basin scale. The net effect is to produce a suite of paraglacial sedimentation curves with varying relaxation times depending on basin scale, i.e. as basin scale increases so will the recovery time. Therefore, in any region where there is a variety of basin scales and there have been periods of asynchronous glaciation, postglacial sediment yields would be expected to vary enormously between basins and over time.

Exploring this idea graphically, one can plot hypothetical sediment yield variations over time for three different basin scales (Figure 6.8). Two important points emerge from this type of plot. First, basins with an identical sediment yield may vary greatly in size, because of thier positions on their respective paraglacial sedimentation curves (for example level S in Figure 6.8). Second, for sediment yield measured at different times, the relative magnitude of sediment yield from basins of different size will vary (e.g. *a* and *b*): at time a the small basin has the highest yield and the large basin the lowest yield, while at time *b* the reverse is true. In this context, if we consider the sediment yield data from Hicks *et at.*, (1990) for the nine schist basins, it is clear that although

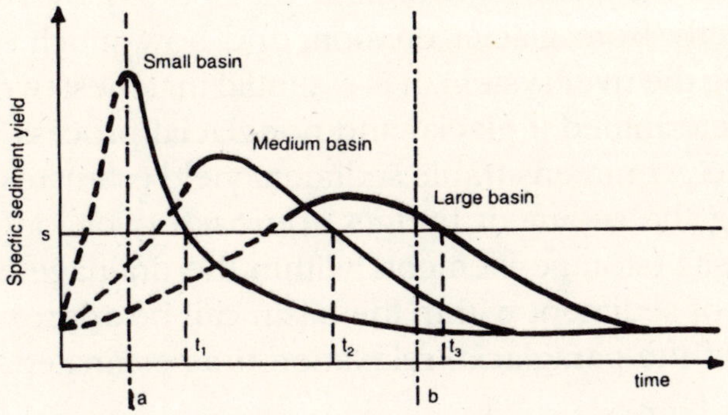

Fig. 6.8. Curves showing paraglacial sedimentation cycles for drainage basins of different scales, given as hyupothetical sediment yield variations over time. Reprinted from Harbor and Warburton (1993).

the Ivory Lake glacierized basin and the nonglacierized basins have similar specific sediment yields, they also have widely different basin scale [the essentially unglacierized basins (< 2 percent glacierized) are all at least eight times larger than the Ivory Lake basin]. What is important is whether the basins have been glaciated or not, the size of the basin, and the lag since the last glacial event. In order to isolate the importance of glaciation on sediment yield a comparison would have to be made between glaciated and unglaciated basins with similar runoff and geology over a variety of spatial scale-an enormous undertaking! (Harbor and Warburton, 1993).

6.7 THE IMPORTANCE OF CONSTRUCTING A GLACIOFLUVIAL SEDIMENT BUDGET

The sediment yield estimates are of limited value in denudation studies because the role of storage in modifying the relationship between erosion and yield is rarely evaluated (denudation estimates based on sediment yields are only strictly valid if change in storage is negligible) (Trimble, 1975; Phillips, 1986). The limitations are involved in using yield data to evaluate the relative magnitudes of glacial and nonglacial processes. A partial solution to these problems would be to construct a detailed sediment budget for each site.

The importance of the sediment budget approach is in quantifying how much of the present sediment transported by river systems is derived directly from glacier erosion, and how much sediment is reworked along the river system. It is essential that thest two quantities are accurately determined if glacial and nonglacial process rates are to be compared, i.e., commensurable sediment yield estimates. Furthermore, by extending the sediment budget approach over longer periods by dating deposits (storage elements within the drainage basin) the residence times of sediment within the basin can be approximated and an assessment of the paraglacial relaxation time estimated.

Borland (1961) provides a relative measure of the intervening effects of sediment storage in proglacial systems by proposing the Borland Index (λ): $\lambda = (A/Ag)\ l$, where Ag is glacier area, A is total catchment area, and l is the length of the glacier stream between the snout and the sediment gauging station. The index attempts to describe the relation between sediment source (expressed as glacier area) and storage, or at least "potential storage" (described by A and l). As the value of the index increases so does the possibility of greater storage in the drainage basin and along the river system. However, stdies relating the Borland Index to sediment yield (Guymon, 1974; Gurnell, 1987) have shown that λ correlates only weakly with suspended sediment yields. Nevertheless in Gurnell's (1987) regression analysis of independent variables used to represent drainage basin characteristics, when considered singly, λ showed the best (negative) correlation with suspended sediment yield.

In the present context, the simple Borland Index can be used to illustrate the importance of storage. For example in the two basins, percent glacier area is identical yet they have very different conditions of possible sediment storage. Thus, percent glacier area only partially defines the important elements in the glaciofluvial sediment transport system. Distance from the glacier to the measurement station is very different in the two cases, and this is critically important as it provides a measure of the potential for storage along the river system as valley fill, a point also emphasized by Ferguson (1984), Gurnell (1987), and Church and Slaymaker (1989).

The Borland Index only partially specifies the main elements of the glaciofluvial sediment system, and to more fully understand the relationship between sediment supply, transport and deposition requires a sediment budget approach (Small, 1987; Warburton, 1990b). Implicit in the discussion so far are four sediment budget models (Figure 6.9); (1) nonglacial, no change in storage; (2) glacial, no change in storage; (3) glacial, change in storage; (4) nonglacial change in storage. Early attempts at estimating denudation in glacierized catchments treated drainage basins as simple black boxes where erosion equalled sediment yield, has largely ignored the effects of changes in storage (models 1 and 2). Growing awareness of the importance of changes in storage has lead

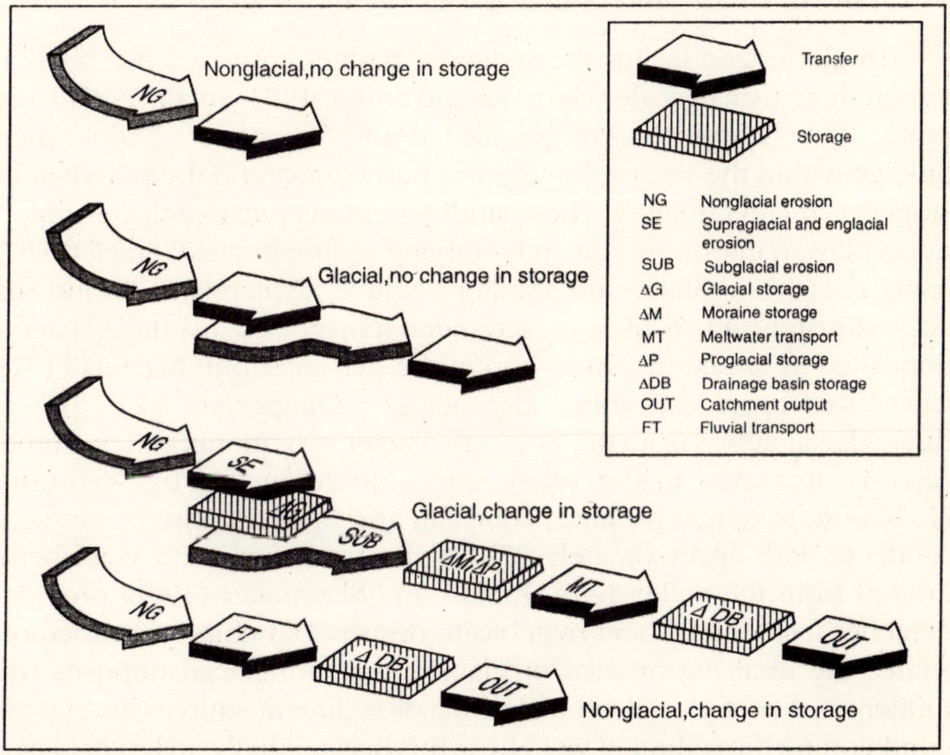

Fig. 6.9. Simple sediment budget models for glacierized and non-glacierized basins. Reprinted from Harbor and Warburton (1993).

to the sediment budget approach (models 3 and 4) where storage changes are explicitly quantified. For example, in their analysis of Ivory

Lake, Hicks *et al.*, (1990) found that, after 1980, the sedimentation rates reduced significantly, averaging 3700 m^3 $year^{-1}$. This reflects the opening of the lower pond which isolated the upper pond from the main sediment sources and caused most sediment to be trapped, undetected, in the deepening in front of the glacier.

From Figure 6.9 (models 3 and 4) it is clear that the main storage elements that need to be evaluated are the glacier, moraines, proglacial zones, and the drainage basin. The inclusion of storage effects is particularly relevant to the study of sediment transfer within glacierized basins where large events can transport the majority of the annual sediment load (Beecroft, 1983) and sediment exhaustion effects occur on seasonal, intermediate, and diurnal cycles (Fenn, 1989).

In glacierized basins the sediment budget approach has been attempted on a small scale (Hammer and Smith, 1983; Small, 1987; Hicks *et al.*, 1990; Warburton, 1990b; Jordan and Slaymaker, 1991), where linkage within the sediment system is fairly direct and the potential for storage is relatively small. These studies demonstrate that these components play an important role in modifying sediment yield. Small (1987), using data from Glacier de Tsidjiore Nouve, Switzerland, found that approximately 40 to 50% of sediment transported by the glacier is deposited in lateral moraines and at the glacier snout. Maizels (1978) found that proglacial zones also act as an important sink, 16% of fluvioglacial debris from the Bossons Glacier, Chamonix valley, French Alps, is deposited in the valley sides. Measurements by Warburton (1990a) indicate that proglical zone can also be an important source of sediment with approximately 25% of the basin sediment yield being eroded from the valley train. Jordan and Slaymaker (1991) provided evidence from the Lillooet river basin, British Columbia, which demonstrates the dual importance of glaciers and nonglacial deposits (the influence of past glacial events) as major sediment sources and factors conditioning the sediment yield from this basin. Finally, in larger basins, far removed from the proglacial zone a considerable amount of sediment is stored within the fluvial system (Ferguson, 1984; Church and Slaymaker, 1989).

6.8 Factors Controlling Suspended Sediment Transport in Himalayan Glacial Meltwaters

The history of sediment measurements in the river Ganga dates back to early 19th century when Rev. R. Everest published his report in Royal Asiatic Society, Bengal, on (soluble and insoluble) sediment transport (Everest, 1832). Contemporary rates of denudation in the Himalaya are high by comparison with continental averages, as might be expected in an area of recent and continuing tectonic activity. The ongoing interaction of the Indian and Eurasian plates maintains uplift, and high elevations ensure high level of precipitation and large glaciers, and steep, unstable slpoes maintain sediment supply to the rivers of the subcontinent (Collins and Hasnain, 1994).

Ageta and Higuchi (1984) categorized Himalayan glaciers as "summer-accumulation" as they receive about 80% accumulation during summer time by the Indian monsoon. Under such conditions summer snowfall sometimes changes to rain under warm air temperature condition, and also new snowcover melts quickly. Consequently, summer air temperature in the Himalaya is an important factor which controls ablation through albedo variation at the glacier surface. All the glaciers in the southern Himalaya are covered with thick supraglacial debris in their ablation area. The debris affects glacier ablation and the runoff amount. It has been observed in the recent years that all the glaciers are receding and forming lakes in their lower ablation regions. Whenever the rainfall portion is increased in total precipitation under warm air temperature, it causes excessive monsoonal rainfall over the accumulation areas of the glaciers and accelerates the melting of new snowfall. Runoff, increased manifold by these processes triggers debris avalanches causing glacier lake outburst and landslides in the downstream. A number of studies have demonstrated, that specific sediment yield may increase downstream, due to remobilization of sediments pushed by active glaciers (Ferguson, 1984; Warburton, 1990). A survey of sediment yield from 1358 drainage basins with an area ranging from 350 to 100,000 km^2 (Jansson, 1988) found that within particular climate zones where glaciers are active, sediment yield tends to be higher.

Rates of sediment erosion in the Himalaya are among the highest in the world. Most of the sediment originates in the glacierized areas and begins its journey to the rivers mixed and pushed by glacier meltwaters. The suspended-sediment transport in the proglacial streams in the Himalayan uplands is controlled by glacial area, glacier activity (retreating), monsoonal rainfall, and changing glacial drainage system (Hasnain, 1996).

6.9 SEDIMENT TRANSPORT FROM BATURA GLACIER, INDUS HEADWATER, KARAKORAM MOUNTAINS, PAKISTAN

6.9.1 Seasonal Variations

The Alpine glacier research group, University of Manchester, United Kingdom, has carried detailed measurements of discharge and sediment flux between April and October 1990 (Collins, 1996). Seasonal variation of daily discharge and measured daily total suspended-sediment transport at Batura glacier is shown in Figure 6.10. Discharge only started to rise in early May, although daily air temperature at the terminus had reached 10°C by mid-April. Most of the glacier remained covered with high-albedo snow so that much of the increasing radiation through May and early June was not utilized in melting. Once snow cover is removed, the lower albedo of underlying ice leads to greatly increased specific yield of meltwater for the same energy input. The first steep rise in flow on 13 May accompanied a rapid temperature increase. Further steep rises in discharge in June, July and August resulted from rapid increase in ice area as the transient line ascended. Rises and falls in sediment flux mimic those of discharge in that when discharge was relatively high so also qualitatively was sediment flux. Sudden increases in suspended sediment transport from a glacier, termed subglacial hydrological events by Collins (1989), result from the interaction of meltwaters with sediment, produced by glacial erosion, that has accumulated at subsole. During minimal winter runoff, products of erosion accumulate across the glacier subsole. As discharges rises, meltwater progressively integrates parts of the subsole that have remained isolated from flow since the previous summer, flushing stored sediment on the first wetting sesaon. Doubling of discharge between 18 and 26 June induced a massive sediment transport event, with 71 kt of suspended material being transferred on 23 June. Sediment flux event magnitudes through July and early August appeared to depend on the marginal increase in wetted area flushed by sediment by progressively higher discharge. Thereafter, sediment flux have been declined with falling flow and shrinkage of basal drainage network.

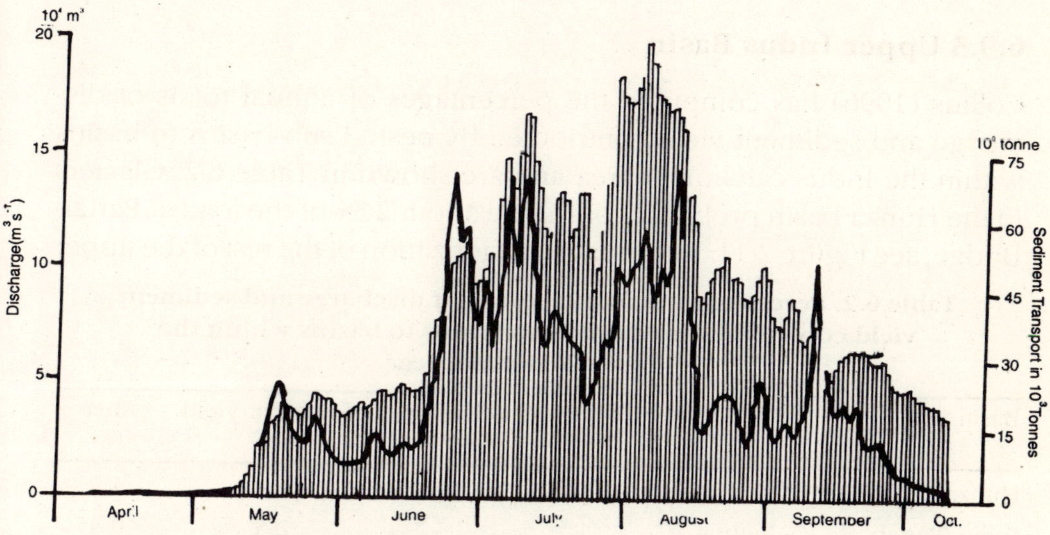

Fig. 6.10. Seasonal variation of daily total discharge (Columns) and measured daily total suspended-sediment transport (Curve) at Batura glacier from April-October 1990, Karakoram, Pakistan. Reprinted from Collins and Hasnain (1995)

6.9.2 Hunza River Basin

The runoff regime of the Hunza river basin at Danyore Bridge reflects the pattern of discharge from the large glaciers in the basin, peaking in July and August. Collins (1996) after analyzing the runoff stated that the 10% of the flow in the Hunza river is derived from the 4.9% of the basin forming the Batura sub-basin. Since there is no monsoonal precipitation in this region, probably the 10% runoff is derived from the 3% of the Hunza basin occupied by Batura glacier. This indicates that glaciers are responsible for the runoff. Mean annual sediment yield for the Hunza at Danyore Bridge is 5kt km^{-1} $year^{-1}$ (Ferguson 1984). The 3762 km2 glacierized area of the Hunza basin might be expected to deliver 38 mt of suspended sediment at Danyore Bridge, which, if one year with measurements at Batura is taken as average, would account for about 60% of the mean annual load indicated by infrequent sampling done by WAPDA, Pakistan.

6.9.3 Upper Indus Basin

Collins (1996) has computed the percentages of annual totals of discharge and sediment yield contributed by nested sub-basins to basins within the Indus catchment area and are shown in Table 6.2. Glaciers in the Hunza basin probably contribute about 24% of the load at Partab Bridge, see Figure 2.11. Percentage glacierization of the rest of the upper

Table 6.2. Percentages of annual totals of discharge and sediment yield contributed by nested sub-basins to basins within the Indus catchment area

Basin	Sub-basin	Discharge% %	Sediment yield %	Area %
Hunza/Danyora	Batura	10.0	6.2	4.9
Indus/Partab	Batura	2.3	2.5	0.45
	Hunza	22.1	39.4	9.2
Indus/Besham	Batura	1.6	–	0.4
	Hunza	16.2	–	8.1
	Indus/Partab	73.0	–	88.3
Chenab/Panjnad	Chenab/Aknoor	26.0	–	12.3
	Jhelum/Baramula	6.8	–	4.5
Indus	Batura	0.7	–	0.2
'mountain	Hunza	6.9	–	3.0
headwaters'	Indus/Partab	31.4	–	32.4
	Indus/Besham	43.1	–	36.7
	Chenab/Aknoor	14.8	–	5.1
	Jhelum/Baramula	3.9	–	2.8
	Chenab/Panjnad	57.1	–	63.3

Indus basin has yet to be obtained. Glaciers on the northern slopes of Nanga Parbat cover 20% of the basin area added between Partab Bridge and Besham and all sediment is glacier produced, 39 kt of sediment will be contributed to the Indus, giving a total flux at Besham of 200 MT year^{-1} (1.2 kt km^{-2} year^{-1}). Of this, the Hunza and Nanga Parbat glaciers contribute about 79 kt or 40% of the suspended load transported by the Indus from the Karakoram and Himalaya. Probably more than 40% of the sediment load leaving the mountains in the Indus is produced by glacial erosion.

6.9.4 Sediment Yields

In the Karakoram mountains, the sediment is mainly derived from the glacier subsole, as the monsoonal rainfall is not penetrating in this region. Therefore, the sediment yield of 10.144 kt km^{-2} year^{-1} is noteworthy (Collins, 1996). Sampled close to the glacier snout the sediment transport in meltwaters will reflect the actual rate of subglacial erosion. High sediment yields from the highly glacierized valleys of the Karakoram indicate the changes in the mass balance, thickness and sliding velocity of the glaciers. For example in the Hunza river basin, glaciers cover about 29% of the area but the subglacial sediment contributions is about 60% and the remainder is presumably derived from the terraces and old moranic deposits left by the retreating glaciers (Collins, 1996).

Sediment transport and yield of the Rakhiot glacier, Karakoram, Pakistan has been studied by Gardner and Jones (1990). According to them total sediment yields are: 4.6 to 6.9 × 10^3 m^3 km^{-2} yr^{-1} for the glacier area, and 1.4 to 2.1 × 10^3 m^3 km^{-2} yr^{-1} for the basin area. These translates into denudation rates of 4.6 to 6.9 mm yr^{-1} for the glacier area and 1.4 to 2.1 mm yr^{-1} for the basin area.

6.10 SEDIMENT TRANSPORT FROM DOKRIANI GLACIER, GANGA HEADWATER, GARHWAL HIMALAYA, INDIA

6.10.1 Climatic Influences

Climatic conditions influencing energy availability for melting and rainfall, and having an impact on both the temporal pattern and quality of runoff, will affect patterns, timing and amounts of sediment delivery from basins distributed along the arc of the Himalayan ranges, (Collins and Hasnain, 1995). Monsoonal cloud cover during the months of July and August reduces energy input for melting, but monsoonal precipitation provides snow at higher elevations and rain at the lower elevations.

Incoming solar radiation with a stronger seasonality at 38° N is intercepted by cloud cover to give periodic depressions in the curve of irradiance of the ground. This leads to sustained reduction in the energy from late June to early August during the monsoon in southern Himalaya. At Batura, on contrary, these reductions are not very pronounced as the monsoonal penetration in the Karakoram mountains is not vigorous as compared to eastern Nepal and Garhwal Himalaya. Therefore, the flows are generally lower in July and August than in June, suggesting that runoff arising from monsoon rainfall and melt of summer snow in the lower zone of the basin more than compensates that lost during periods when newly fallen snow prevents melting of ice throughout the elevation range. The difference between regimes of runoff from the drier west and monsoonal central and eastern Himalaya probably leads to contrasting patterns of sediment flux along the Himalayan arc.

6.10.2 Diurnal Variations

Diurnal variations in discharge and suspended sediment concentration are characteristics of glacial-fed rivers, because melt processes is directly related to the diurnal variations in radiative energy input (Collins and Young, 1981). Diurnal variations in the suspended sediment concentration (C)-discharge (Q) relationship of the Dokriani glacier shows that the relationship varies through different stages of ablation season. Figure 6.12 shows the diurnal discharge-suspended sediment realtionship July 27, August 18 and September 7, 1994. From the last week of July to early September, gradual strengthening in the mean relationship between suspended-sediment concentration and discharge was observed. Discharge as well as suspended-sediment concentration peaks became progressively earlier in the day, resulting in short rising limb and extended falling limb. These characteristics probably result due to variations in glacier melting and precipitation, water discharge in the subglacial conduits is subject to changes throughout ablation season (Bogen, 1996).

6.10.3 Suspended Sediment Transport

Figure 6.11 shows seasonal variations of daily suspended-sediment transport and daily total discharge for 1994 ablation period. From May 19 onwards discharge started to increase and was associated with an increased sediment transport. In June, rise of the transient snowline led

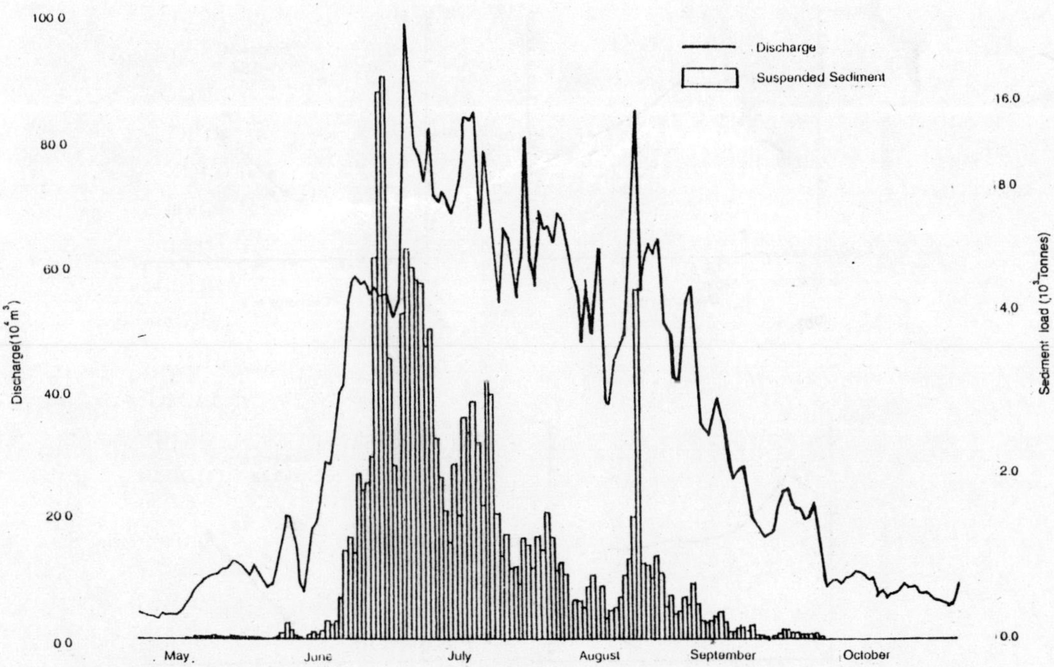

Fig. 6.11. Seasonal variation of daily total discharge (curve) and measured daily total suspended-sediment transport (columns) at Dokraini glacier from May-November 1994, Garhwal Himalaya. Reprinted from Hasnain and Thayyen (1999b)

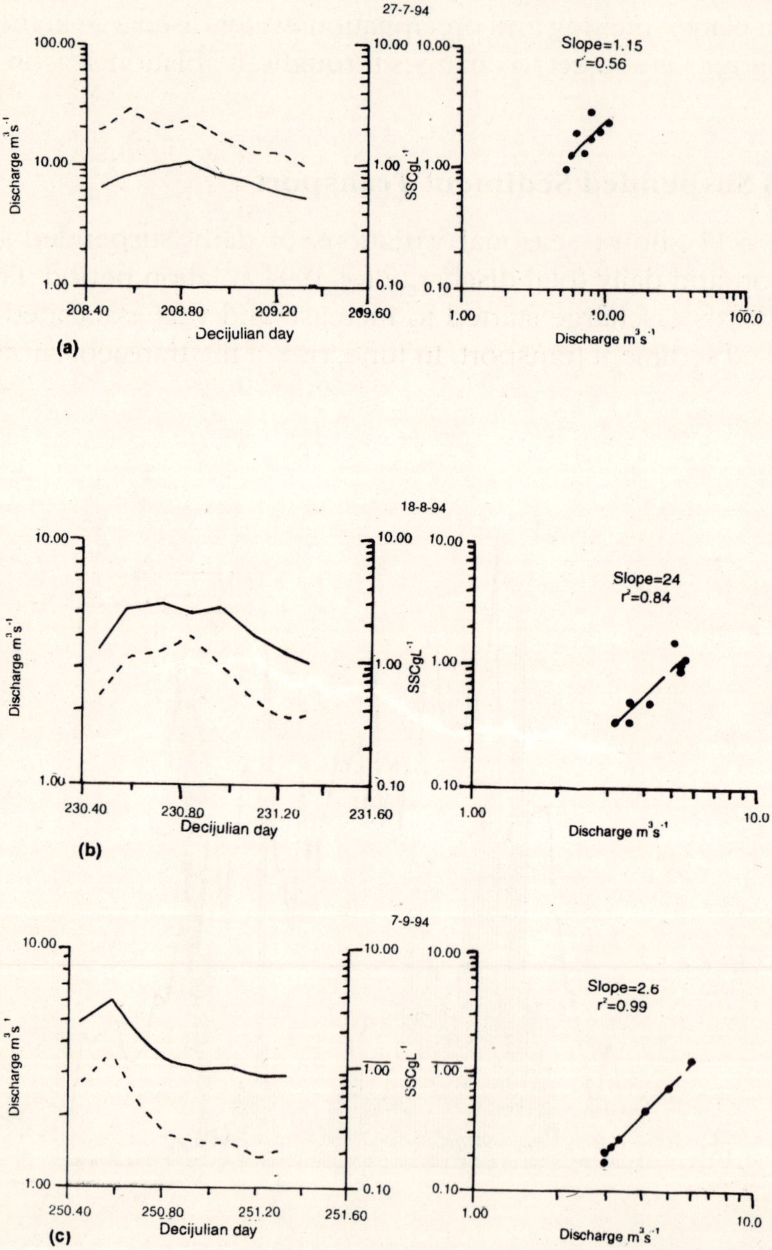

Fig. 6.12. Relationship between diurnal discharge and suspended-sediment concentration in July, August and September in the Dokriani glacier, Garhwal Himalaya. Reprinted from Hasnain and Thayyen (1999b)

to discharge doubling with sediment transport of 4×10^4 tonnes. This is compatible with the concept developed by Collins (1989) that the areas of previously hydraulically isolated sub-sole are integrated with flow, releasing large quantities of sediment from basal storage. Most of the Alpine glaciers of the world show sediment exhaustion and failed to rise to earlier levels in spite of high flows in July and August (Collins and Hasnain, 1995). Whereas, in the Dokriani glacier basin sediment transport rises to 8×10^4 tonnes in July as shown in Table 6.3. One possible explanation follows: continued subglacial sediment evacuation and onset of monsoonal precipitation. This suggests that the high intensity monsoonal rainfall over the ablation region pushing surfacial cover of moraine debris (supraglacial moranie) through extensive longitudinal and transverse crevasses into englacial and subglacial tunnels and causing the high sediment transport in meltwaters. However, September and October show marked recession of both fluxes, as a result of fresh snow fallen on the accumulation region of the basin.

Table 6.3. Monthly discharge, sediment flux and rainfall during the 1994 ablation period

Month	Total discharge (10^4 m^3)	% Discharge	Sediment flux (10^4 tonnes)	% Sediment flux	Rainfall (mm)	% Rainfall
May	195.5	3.0	.04	.028	4.0	
June	868.0	14.0	4.0	26.0	15.0	9.0
July	2187.0	35.0	8.0	52.0	500.0	39.0
August	1785.0	29.0	3.0	18.0	434.0	34.0
September	910.0	14.0	0.52	3.5	189.0	15.0
October	292.0	5.0	0.02	0.19	26.0	2.0
Total	6238.0		15.0		1270.0	

Monthly contrasts in the relationship between suspended-sediment concentration and discharge are shown in Figure 6.13. In July there appears to be little association between suspended-sediment concentration and discharge. This probably is caused by the high intensity monsoonal rainfall (600 mm), which disturbs the seasonal pattern of sediment flux from glaciers. Hence, the climatic conditions have a substantial

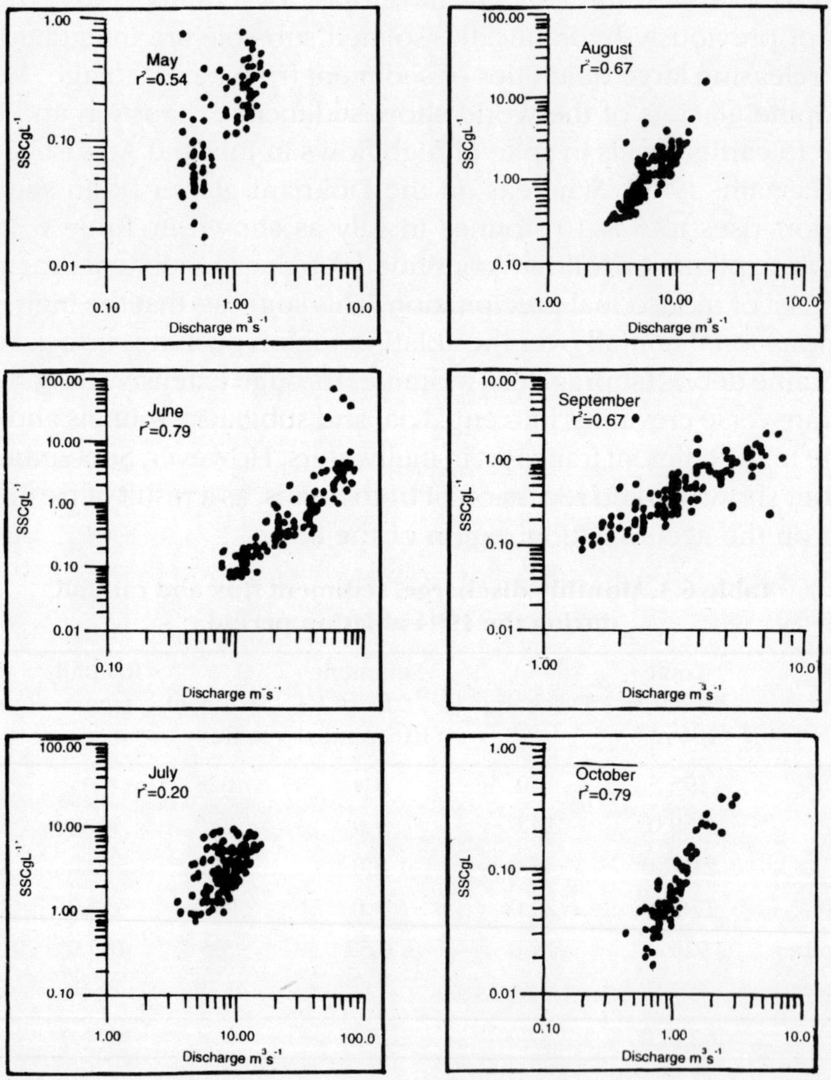

Fig. 6.13. Monthly varaitions in the relationship between suspended-sediment concentration and discharge for the months of May, June, July, August, September and October 1994. Reprinted from Hasnain and Thayyen (1999b)

influence on sediment evacuation (Collins and Hasnain, 1995). During the ablation period a total of 15×10^4 tonnes suspended-sediment was transported from the Dokraini glacier basin. 52% of the total ablation period sediment transport occurred in the month of July (i.e. 8×10^4 tonnes) alongwith the highest volume of discharge (35%). The months of July and August (monsoonal) contributed 64% of the total discharge, 70% of the total suspended transport and experience 74% of rainfall (934 mm). During June, 26% of the total suspended-sediment was transported, though the discharge was less than that in August. This is probably due to the evacuation of preceding summer/winter sediments from the subglacial region.

The suspended sediment transport-discharge relationship shows seasonal and diurnal variations in Dokriani glacier meltwaters as shown in Table 6.3. On the seasonal time-scale, the sediment transport rises to 8×10^4 tonnes in July from 4×10^4 tonnes in June, an 100% increase caused by the impact of monsoonal precipitation and not subglacial sediment evacuation. During the six months ablation period, the monsoonal months of July and August account 64% of the discharge, 70% of the suspended sediment transport, and 74% of monsoonal rainfall. On the diurnal time-scale discharge and suspended-sediment concentration peaks became progressively earlier in the day, resulting in short rising limb and extended falling limb. These characteristics are probably linked with variations in the glacier ice melting caused by the monsoonal regime over the Ganga basin glaciers resulting in the liquid precipitation on the ablation area and the solid precipitation on the accumulation area simultaneously. The results clearly indicate that regional climatic conditions perhaps influence energy availability for melting, length of monsoon season and quantity of precipitation and all these variables are controlling the amount of sediment delivery from glacierized basins. Table 6.4 gives comparative sediment transport data from glacier in Karakoram, Pakistan, eastern Nepal and Garhwal Himalaya, India respectively. The difference in runoff regime between non-monsoonal Karakoram and monsoonal Himalayan catchments (Collins and Hasnain, 1995) therefore probably leads to contrasting patterns of sediment flux.

Table 6.4. Annual sediment transport data for three important Himalayan glaciers.

Glacier	Area ($10^3 km^3$)	Mean annual discharge km^3	Glacierization %	Mean annual transport (Mt)(tkm^2yr^{-1})	Source
Batura, Karakoram, Pakistan	0.64	0.93	60.0	4.0; 6068	Lanzhou Institute, 1984
Langtang, Nepal	0.33	0.45	38.0	.08; 243.0	Fukushima *et. al.*, 1987 Ohta *et at.*, 1987
Dokriani Garhwal Himalaya, India	0.0096	.06	60.0	0.15; 15751	Hasnain (1996)

7

GLACIER LAKES AND THEIR OUTBURST FLOODS

7.1 INTRODUCTION

The great Himalaya provides a sub-polar climate and an appropriate condition for a stable existence of cryosphere covered by snow and glaciers. Presently glaciers exist in the areas above, 4,000 m. Some of these glaciers have a glacial lakes. The formation of such lakes seems to have accelerated in recent times (Yamada, 1998). Many glacier lakes were born during the last half century in the Himalaya. The recent information of glacier lakes is believed to be visual evidence of the global warming.

All the glaciers in the Himalaya and Karakoram mountains are in a state of retreat since the middle of the 19th century. In the process of a glacier retreat, ice tends to melt in the lower part of the glacier surrounded by lateral and end moraines and a pond/ponds is made up of this meltwater as shown in Figure 7.1. The structure of the storage is not consolidated and invariably not stable. A slight disturbance can easily change the balance of the lake and an abrupt release of great amount of stored water may result in and cause a flood. Serous damages are caused to infrastructure, to the inhabitants and also eco-system and environment downstream in the Himalayan regions. This phenomenon is called glacier lakes outburst floods (GLOF).

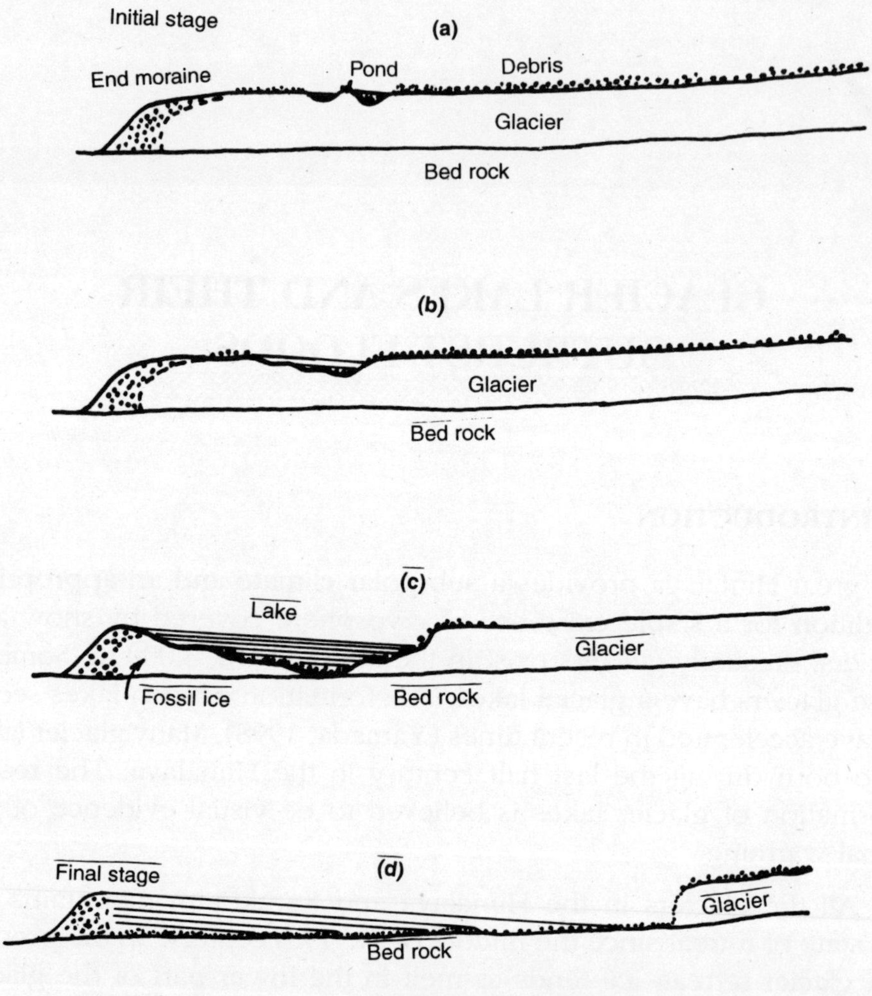

Fig.7.1. Development of a moraine-dammed glacier lake. Reprinted from Yamada (1997)

7.2 GLACIER LAKES

Two kinds of glacier lakes exist in the Himalaya, one is moraine-dammed lakes and the second ice-dammed lakes, as shown in Figure 7.2. Himalayan glaciers produce very large amount of debris and have large lateral and end moraines compared with other glaciers in the world. The glacier tongue is covered by thick debris and contains debris in the glacier body. The terminus area of tongue does not cause flowing action and may remain stagnant as a dead ice mass. This area is separated from the dynamic mass balance system of the presently active glacier. Its real terminus is now located in the upper part, though this area looks apparently connected to the present real terminus to the glacier.

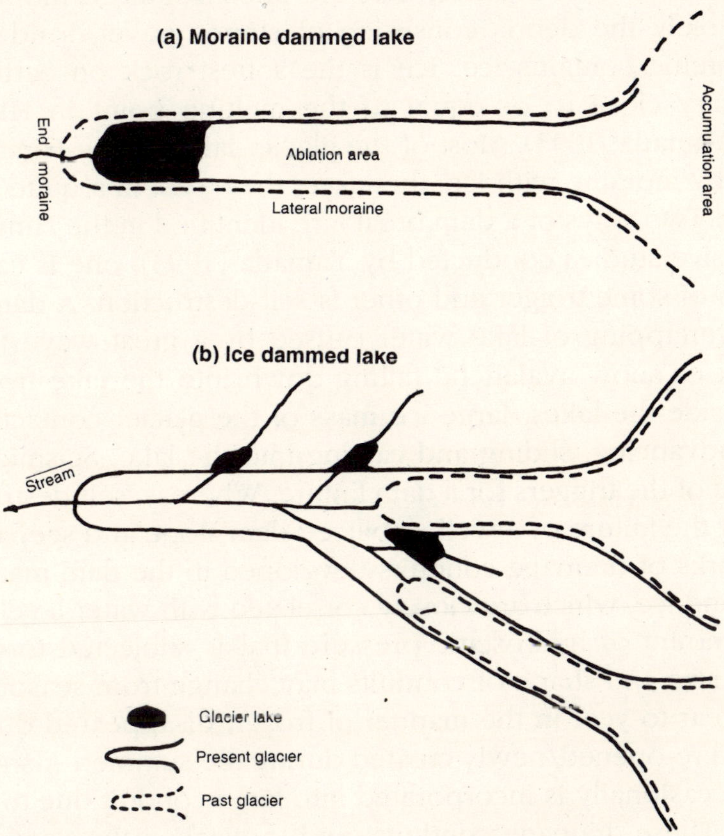

Fig. 7.2. (a) moraine dammed lake, (b) Ice dammed lake. Reprinted from Yamada (1997)

All the lakes in the Himalaya and Karakoram are moraine-dammed in origin and ice dammed lake is rare. A glacier lake my disappear, once the dam is destroyed or sedimentation fills up the lake completely. It is only formed and maintained in certain stages of glacier activities corresponding to climatic variation. According to Yamada (1993), a glacier lake forms and disappear according to the activity of the glacier. Existence of a glacier lake is essentially temporary and not stable from the point of view of the glacier's history.

7.3 CAUSES OF LAKE BURST

The dam materials of a glacier lake are unconsolidated moraine. Moraine is solely the debris consisting of stone, gravel, sand and silt. Some moraine contains ice. Ice is the softest rock on earth and is usually very close to or equal to the melting point in Himalayan regions (Yamada, 1993). Most of the glacier lakes in the Himalaya are dammed by moraine with ice; therefore, they exist in a quite unstable condition. Two ways of a dam break are identified in the Himalaya by the extensive studies conducted by Yamada (1993); one is caused by the action of some trigger and other is self-destruction. A dam breaks by the overtapping of lake water caused by a great wave triggered from rock or snow avalanche falling down into the lake from steep slopes beside the lakes, large ice mass of the glacier contacting with the lake advancing, sliding and calving into the lake. Seismic activity also is one of the triggers for a dam failure. Whereas, self destruction is caused by the failure of a well stabilized dam slope and seepage from the networks of drainage conduits developed in the dam materials of moraine and ice, which are closely correlated with water level of lake, i.e. the amount of hydrostatic pressure that is subjected to the dam wall. The size and shape of conduits may change from season to season and year to year in the manner of frozen/disappeared during the winter and re-opened/newly created during the summer. Also, a small conduit occasionally is incorporated into large conduit due to erosion by water. These drainage conduits tend to weaken the dam collapse due to water pressure stored in the glacier lake without any particular trigger.

As a result, huge amounts of water from the glacier lake are suddenly released and rush down along the river channel heading downstream area as a sharp flood wave, which results in the washing away of all the infrastructure along the river.

7.4 GLOF IN THE KARAKORAM

Thirty-five destructive outburst floods have been recorded (Table 7.1) in the past three hundred years in the upper Indus basin, Karakoram Himalaya, Pakistan has been reported by Hewitt (1982). Glacier dams in the main river valleys have occurred from the far western to the far eastern parts of the Karakoram ranges as shown in Figure 7.3. According to Hewitt (1982), the main Indus river follows extremely elongated, trellis drainage patterns that divide up the high ranges. The high ranges of Karakoram are heavily glacierized and have 60 to 70 % permanent snow and ice covers. Snowfall of the order of 1000-1500 mm water-equivalent is present here.

Table 7.1. Historical summary of glacier dams and outburst floods for the Karakoram Himalaya, (after Hewitt, 1982).

Year	Ice Dam / Moraine Dam	Glacial / River System
1533	×	Upper Shyok river
1780	×	Upper Shyok river
1826	×	Upper Shyok river
1833	×	Upper Shyok river
1833	×	Yashkuk Yaz Glacier
1835	×	Sultan Chhussku river
1842	×	Upper Shyok river
1844	×	Ishkoman river
1848	×	Aktash Glacier
1848	×	Kichik Khumdan Glacier
1850(?)	×	Chungphar Glacier
1850	×	Aktash Glacier
1852-1858	×	Kichik Khumdan Glacier
1855	×	Upper Shyok river
1855	×	Upper Shyok river
1864	×	Kichik Khumdan Glacier(?)
1865	×	Ishkoman river
1869-1872	×	Kichik Khumdan
1870	×	Karambar Glacier
1873	× (?)	Batura Glacier

***Table 7.1** (contd.)*

Table 7.1. *(contd.)*

Year	Ice Dam / Moraine Dam	Glacial / River System
1879	×	Upper Shyok river
1882	×	Upper Shyok river
1884	×	Shimshall river
1889	×	Upper Shyok river
1891-1892	×	Ishkoman river
1893	×	Shimshall river
1893	×	Ishkoman river
1899	× (?)	Upper Shyok river
1901	×	Upper Shyok river
1902-1911	×	Kichik Khumdan Glacier
1903	×	Kichik Khumdan Glacier
1904-1905	×	Ishkoman river
1905-1906	×	Khurdopin Glacier
1905	×	Kichik Khumdan Glacier
1905	×	Karambar Glacier
1905	×	Khurdopin Glacier
1906	×	Khurdopin Glacier
1907	×	Khurdopin Glacier
1909	×	Whirgut Glacier
1916	×	Ishkoman river
1924-1933	×	Chong Khumdan Glacier
1925-1927	×	Khurdopin Glacier
1926	×	Chong Khumdan Glacier
1927	×	Kaz Yaz Glacier
1927	×	Khurdopin Glacier
1928	×	Kilik river (Hunza)
1929	×	Chong Khumdan Glacier
1929-1930	×	Ishkoman river
1930	×	Kyagar Glacier
1932	×	Chong Khumdan Glacier
1933	×	Chong Khumdan Glacier
1939	×	Chong Khumdan Glacier
1953	×	Kutiah Glacier
1978	×	Kyagar Glacier
1979	×	Kyagar Glacier

Fig. 7.3. Distribution of glacier dams in the Upper Indus basin. Reprinted from Hewitt (1982)

7.4.1 Outburst Floods

The 1929 outburst flood of Chong Khumdan glacier was monitored from near the glacier over more than 1500 km downstream (Gunn, 1930; Mason,*et al.*, 1930). Along with some comparative observations for the smaller 1932 outburst, (Mason, 1932) this gives a past record of flood-wave behaviour on the upper Indus. Gunn (1930) estimated the reservoir to have contained almost 13.5×10^8 m^3. Some 3×10^5 m^3 of ice were also carried with the flood and stranded on large blocks in the valley below the dam. If loss to the channel storage and from seepage is somewhat greater than gains from inflows below the dam, the complete drainage in 48 h suggests an average discharge between Sasir Brangsa and Khalsar in the region of 7100 m^3 s^{-1}. That equals the larger discharges measured for the entire upper Indus at Attock.

The little information available of the 1926, 1929 and 1932 Khumdan outbursts is that breaching began through subglacial tunnels, but then carried away the entire thickness of the ice above.

7.4.2 Sediment Transfer

Sediment yield from the upper Indus basin represents the highest known rate of regional erosion, of about 1 m per thousand years. Data for the tributaries where the dams occur suggest rates in excess of 1.8 m per thousand years. High sediment concentration generally occurs in association with flood waves as shown in Figure 7.4.

The great height and erosional energy of dam-burst flood waves especially allows their to reach and cut into the abundant lag sediment deposits in or stranded upon, arid river terraces. Landslides may occur on terraces and valley sides after the passage of damburst floods (Hewitt 1982).

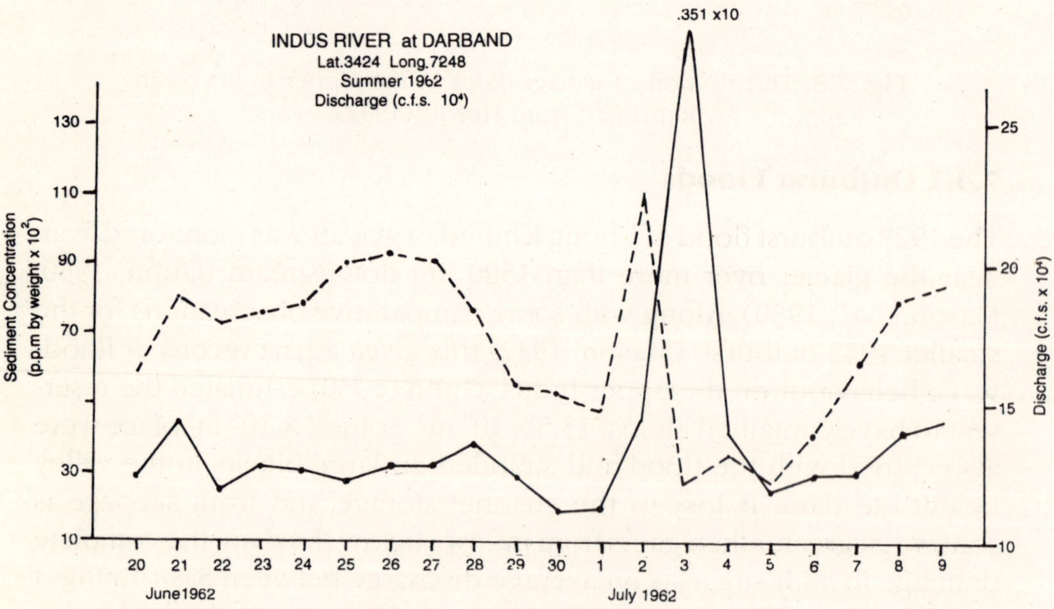

Fig. 7.4. High sediment concentration (broken line) in association with flood wave (solid line) on the Upper Indus river. Reprinted from Hewitt (1982)

7.5 GLOF IN THE NEPAL HIMALAYA.

The GLOF studies were made in glacierized areas in Nepal and Tibet by Yamada, (1998). The past GLOF's were studied by means of satellite imageries, field visits to glacier lakes and collection of GLOF information from the inhabitants living in the Himalayan highlands along the rivers. Table 7.2 shows a list of GLOF events which happened in the past in Nepal. GLOF in Nepal Himalaya mainly happens from moraine-dammed lakes. The peak discharge of this type of flood is sharper as compared to ice-dammed lakes. The high power of rushing water destroys banks of rivers and causes a collapse and landslides from steep slopes along river channel. As a result flood water includes a great amount of debris. There are a number of small ponds on debris-covered glaciers in the Nepal Himalaya, which may be the embryo of a glacier lake. Some of them have often burst and have caused small floods.

7.6 GLOF IN THE INDIAN HIMALAYA

Little is known about the GLOF in the Indian Himalaya. There were many glacier lakes formed in the vicinity of the major glaciers in Kasling Himachal Pradesh, Uttar Pradesh and NE Himalaya. Some of them have already burst out causing extensive damage in downstream valleys. Nevertheless, many of the floods in the Ganga headwaters were caused by the GLOF. Unfortunately, there is no data and studies on the GLOF in the Indian Himalaya.

Historical records reveal numerous floods in the Alaknanda valley. Atkinson (1881) has documented some of the sudden floods in the Garhwal Himalaya. The floods of 1803, 1868, 1880,1893 were GLOF from various glaciers feeding the Alaknanda and its tributaries.

Walker and Pacoe (1907) described the existence of a very large glacier lake in front of Bara Shigri glacier snout and the photograph taken by them also shows the presence of huge boulders in front of snout. However, the morainic dam of the lake was effected either by a flood due to intense melting of glacier system upstream of Chandra river or by the overflow of the lake caused by a landslide. The studies conducted by Dutt (1961) described the absence of large boulders and

Table 7.2. : List of glacier outburst floods happened in Nepal, Himalaya.

No	Date	River basin	Name of lake	Cause of GLOF	Remarks and Damages
1.	450 yrs ago	Seti Khola	Machhapuchhare	Ice-cored moraine collapse	Pokhara covered by 50 - 60 m thick debris
2.	Aug. 1935	Sun Kosi (Targyailing Gully)	Taraco	Moraine collapse due to seepage	66,700 m^2 of wheat field, livestock, etc.
3.	21 Sep. 1964	Arun	Gelhaipco	Moraine collapse due to glacier sliding into the lake	Dam breach 40 m deep and 23 million m^3 of water draining out. Bulk density of 1.45 t/m^3 highway and 12 trucks etc.
4.	1964	Sun Kosi	Zhangzangbo	Moraine collapse due to seepage	No remarkable damages
5.	1964	Trisuli	Longda		
6.	1968	Arun (Zongboxan river)	Ayaco	The lake bursting three times in 1968,1969,and 1970	Road, concrete bridge etc.
7.	1969	Arun	Ayaco		
8.	1970	Arun	Ayaco		
9.	3 Sep. 1977	Dudh Kosi	Nare	Ice-cored moraine collapse	Mini-hydro plant, road, bridges, farm land
10.	1980	Tamur	Punchan		One village destroyed; villagers migrated after GLOF

Table 7.2 *(Contd.)*

Table 7.2 *(Contd.)*

No	Date	River basin	Name of lake	Cause of GLOF	Remarks and Damages
11.	11 July 1981	Sun Kosi	Zhangzangbo	Moraine collapse due to glacier falling into the lake	Dam breach 50 m deep and 40 - 60 m in width; max drainage being 16,000 m^3; 19 million m^3 of water drained out. Arniko Highway, friendship bridge, Sun Kosi hydropower station, farm land etc.
12.	27 Aug. 1982	Arun	Jinco	Moraine dam collapse due to glacier ice avalanche	1600 heads of livestock, 187,000 m^2 of farm land, houses of eight villages washed away
13.	4 Aug, 1985	Dudh Kosi	Dig Tsho	Moraine collapse due to glacier ice avalanche	Namche hydropower station, 14 bridges, trails, cultivated land etc.
14.	12 July 1991	Tama Kosi	Chubung	Moraine dam collapse of Ripimo Shar glacier	Houses, farm land etc.
15.	May 1995	Kali Gandaki	?	Tsarang Chu upstream in Mustang	

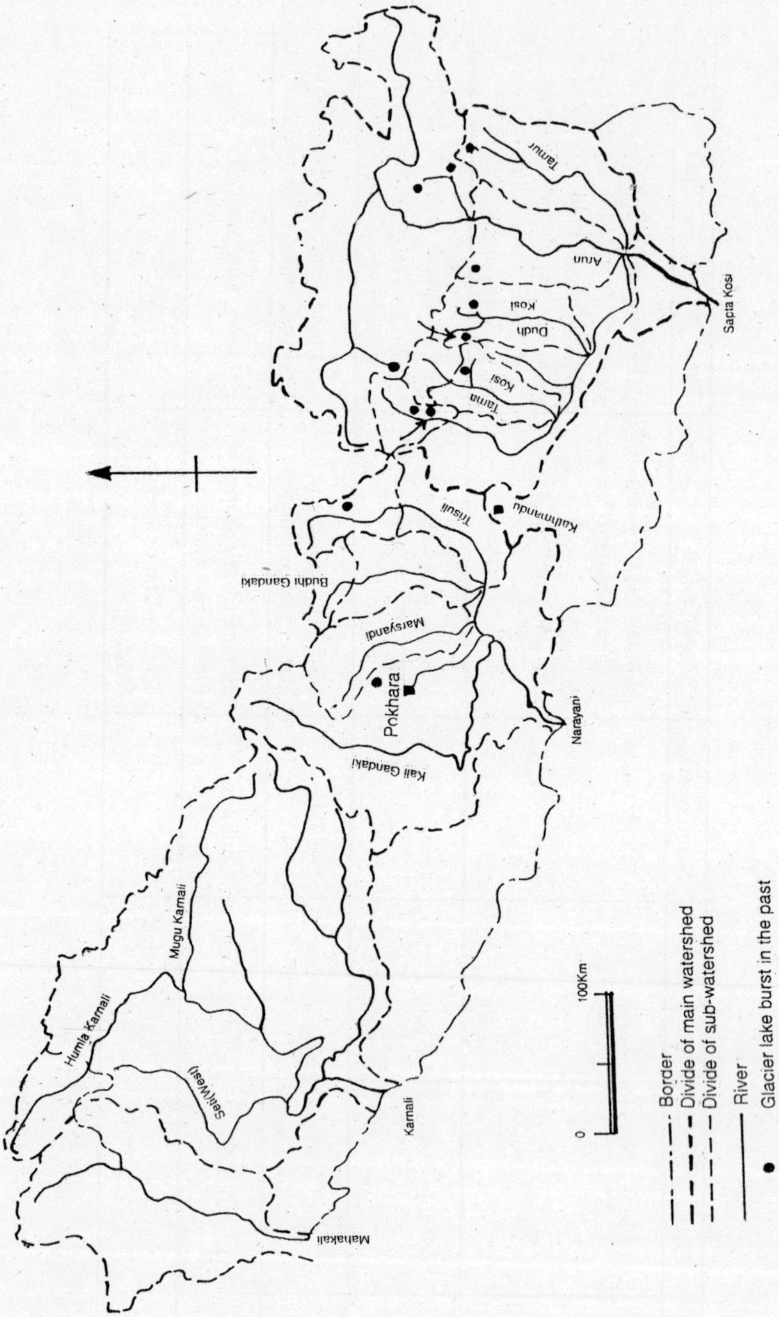

Fig. 7.5. Distribution of glacier lakes that burst in the past in Nepal. Reprinted from Yamada (1998)

the lake. Perhaps all the boulders must have disappeared by the surging waters of the lake.

A visit to glaciers in the Alaknanda valley indicates the existence of glacier lakes near their vicinity. The field observations clearly indicate the existence of glacier lakes near some of the major glaciers. The Khatling glacier, feeding to the Bhilagana river, one of the tributaries of Alakananda river shows that a moraine-dammed glacier lake was collapsed in 1970's causing devastating flood in the valley. Similar lakes were formed near the glaciers in Dhauliganaga, Mandakni, Nandakni, Patalganga, Birahiganga, all are tributaries of Alaknanda river and they are heavily galcierized in their headwaters. They all suffered GLOF during the recent and ancient times causing major floods/debris avalanches.

The GLOF have also been causing floods in the Bhagirathi river. The headwaters of Bhagirathi and its other tributaries is heavily glacierized as shown in Figure 2.5 in Chapter 2. The floods of 1978 and earlier floods were related to the glacier lake outburst. These days a very big glacier lake is developing on Satopanth glacier — one of the major feeding glacier to the Alaknanda river. The GLOF studies in the Indian Himalaya are not properly catalogued.

APPENDIX I

1 GLACIER MELTWATERS MEASUREMENT

1.1 General

The water draining from a glacierized basin originates from mainly four different sources.

1) Glacier melt

2) Rain falling on the glacier

3) Rain falling on the glacier-free parts of the basin

4) Inflow of ground water

Apart from some water loss to the atmosphere through evaporation (or water gained by condensation in humid areaa), the sum of the measured ablation on the glacier and the total volume of precipitation will be discharged in the glacier stream. To check short-term ablation measurements, it is necessary to measure the water discharge as accurately as possible.

Such gauging stations provide a unique, long-term record from a glaciated basin where complementary long-term records of glacier topography and mass balance fluctuations are obtained. This type of information is essential for glacier hydrologic studies and runoff modeling. Proper physically-based models of glacierized areas can only be developed if high-quality runoff data are available over a long time period, and if glacier accumulation and melt studies are conducted at the same time.

1.1 RATING CURVE AND NUMBER OF READINGS

Discharge calculations are generally based upon water-level readings on a vertical gauge or on data from an automatic water-level recorder. The relationship between the water level at a given site and the actual discharge (expressed in $m^3 s^{-1}$) must be determined by many direct measurements of river flow for different gauge readings to obtain a graph of level vs discharge, the rating curve (Figure 1A.1), for each measuring site.

Appendix-I

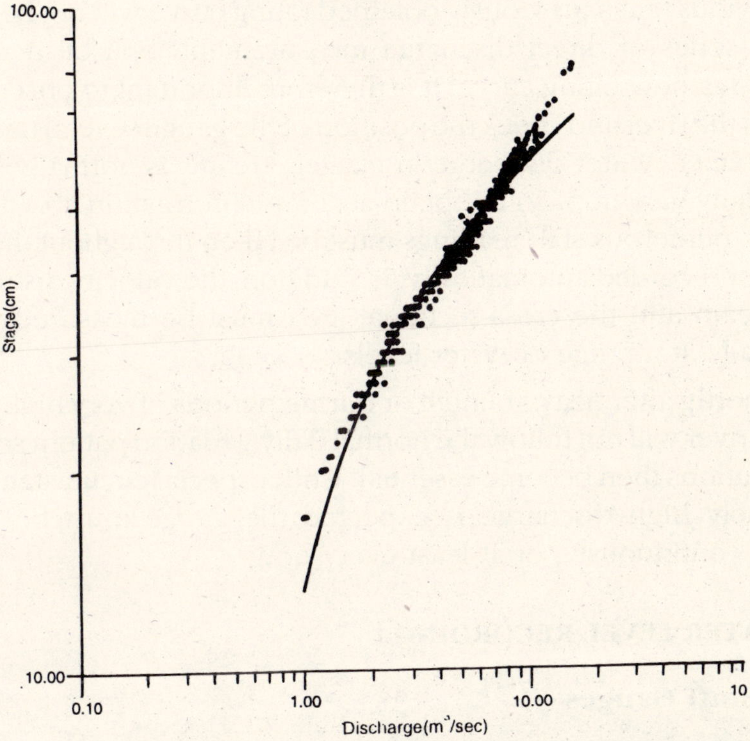

Fig. 1A.1. Ratuis curve of glacial meltwaters in the Himalaya

It is assumed that the river discharge conditions will not be disturbed by changes in the river bed. Ideally, the cross-sectional area should be controlled by bedrock and the vertical gauge placed in a little pool above this control. If the channel changes, the rating curve is no longer valid and a new cvrve must be constructed, i.e. a new series of direct discharge measurements must be made.

Direct measurements of water discharge can be obtained in many ways. For streams with "laminar flow", a standard current meter can be used to measure water velocity, but this method is difficult to use in glacier streams because they are often highly turbulent. Turbulent streams can be measured using calorimetric or fluorometric methods.

Once a rating curve has been established all future gauge readings can be easily transformed to discharge figures. However, if the river cross section changes or the gauge is moved relatively to the river bed, it is obvious that a previously obtained rating curve will not be valid and a new series of direct discharge measurements will be necessary to establish a new rating curve. It is therefore important to observe conditions in the river and check the position of the gauge several times during the summer. Water discharge in glacier streams is normally subject to great daily variation, so to obtain accurate information about total discharge, numerous staff readings must be taken throughout the day and night, or recorded automatically. In addition, the velocity distribution of the stream and the cross-sectional area must be measured regularly, especially at a range of water levels.

Shortly after heavy rainfall or during periods of very high temperatures a river will not follow the normal daily variation pattern so frequent observations then become essential. Under special circumstances when extremely high discharge is expected the gauge must be observed almost continuously, or at least every hour.

1.3 WATER LEVEL RECORDING

1.3.1 Staff Gauges

A gauge is a vertical rod marked in meters and centimeters and placed so that the height of the water level can be measured. The gauge should be placed in the river where the water is as tranquil as possible. A pool in bedrock may be the best site to erect a gauge, but it must be located so that both a small discharge and a very large discharge can be measured directly.

In the fall, or when the water discharge is very low, it is necessary to determine a reading on the gauge that corresponds to zero water discharge (i.e. generally the height of the threshold in the pool). This reading gives important information for construction of the rating curve. A staff gauge is required for calibration purposes, even if an automatic gauge is installed.

1.3.2 Automatic Gauges for Stage Recording

To reduce the time-consuming observation programme described above, an automatic gauge can be installed. In principal, the original automatic gauge consists of a float, a counter-weight and recording device, and these are still used in many locations. Variations in water level move the float and the counter-weight and the movement is transferred to a pen that registers the movement on a paper-wrapped drum. The drum is driven by a clock and one rotation may be completed in 24 hours, one week or a month. For glaciological use, one must select the most suitable period related to the frequency of summer visits to the glacier.

Every time the chart is changed or data retrieved from the data logger, and, if possible, once or twice in between, the water level, as read on the vertical gauge should be written directly on the chart together with information about date and time and noted. This check is important when the record is used for calculating total water discharge.

As the records from an automatic gauge give information about water level(m) it is necessary to convert the data into discharge figures ($m^3\ s^{-1}$) before further calculations are made.

1.3.3 Gauging Methods

Several methods can be used for monitoring stream stage, which must then be related to stream discharge. This is done through a rating curve mentioned earlier. Both a float gauge and an acoustical sounder have been used.

Bubble gauges and pressure transducers are also alternatives for measuring water levels, they are largely used in Switzerland and France for measurements on glacier meltwaters. As data loggers become more affordable, they should provide the best method for recording the data. Telemetry is also becoming more affordable whereby data can be sent using relatively inexpensive radios connected to phone jacks. The data should always be stored in the logger as the field site in the telemetered data is somehow lost during transmission.

1.3.4 Calculation of Discharge Water Volume

Calculations of water volume of the discharge during a given time interval can only be made if a rating curve has been established. From the readings on the gauge (the vertical scale), from the automatic gauge chart, or the data logger file, it will be possible to calculate the water discharge for any period. If the water level shows rapid fluctuations within short time intervals, it may be necessary to calculate the discharge every hour or at even shorter intervals. When the water level shows small variations, i.e. when the discharge is fairly constant, periods of several hours can be used for the calculation. Generally a 6-h period will suffice.

For each selected period, a mean water level should be determined. The corresponding water discharge is found from the rating curve or from a nomogram. It is anticipated that it will be representative of the period and the total volume of water can be found by simple multiplication.

When necessary the volume of water discharge should preferably be calculated for the following 6-h periods: 0000-0600, 0600-1200, 1200-1800 and 1800-2400 (midnight). The greatest discharge will probably occur during the last two periods so special attention should be given to possible variations within them. The results of the calculations should be recorded on summary forms. The sum of the discharge water volumes in the above-mentioned 6-h periods is called the daily discharge. Variations in daily discharge may later be compared with meteorological parameters, such as daily mean temperature, daily precipitation.

1.3.5 Evaluation of Discharge Measurements Methods in the Himalaya

All the rivers in the Himalaya are very steep, seasonally shallow, fully turbulent with extreme roughness coefficient due to the many huge boulders in the water course and the many hydraulic jumps. The cross-sections cannot be obtained with accuracy, partly because of the inaccessibility of the river banks on both sides (no bridges), partly because of the blocky nature of the river cross-sections that tend to change. During high floods, bed-load transport of pebbles and bigger stones is common.

The following discharge methods have been evaluated:

1. Flumes and Weirs:

Considered to be very accurate method, but the cost and logistics of establishment in these areas is a daunting task. The meltwater rivers like Bhagirathi and Alaknanda, in the Garhwal Himalaya and Chandra river, in the Lahul Spiti valley require large scale upstream construction work. The movement of boulders and bed load, siltation, damage to structures after floods make it difficult to operate over a long period of time.

2. Current meters:

A principal pre-condition for this method is that the cross-section is known with sufficient accuracy. Bridges to suspend current meters are almost absent. The unstable hydraulic conditions, high flow velocities, bedload transport are some of the limiting factors for the use of the current meters.

3. Salt Dilution:

The major limiting factor in the use of this method is large quantity of salt required and its transport for the discharge site.

4. Tracer method:

Dye tracers have been used for over a century to investigate glacial drainage system. Fluorescent dyes such as rhodamine or fluorocein are the most commonly used tracers, since they can be detected by fluorometer at concentrations of less than 1 ppb upon emergence in the proglacial stream.

The mean water throughflow velocity during a test can be obtained from the time elapsed between dye injection and peak concentration at the detection site. The flow velocity through the subglacial drainage system is dependent on the character of the flow pathways via which the water has travelled. Rapid transit velocities of greater than 0.2 ms^{-1} are commonly interpreted as indicative flow through hydraulically efficient drainage systems (Lang *et al.*, 1979) In contrast, lower throughflow velocities have usually been interpreted in terms of flow through less efficient hydrological configurations

Theakstone and Knudsen, (1981) found to be suitable in the Himalayan upland rivers. It is currently used in Nepal Himalaya for measurements of discharges. The method is ideal to be used in steep, fully turbulent rivers with discharges of under 100 $m^3\ s^{-1}$ and where cross section measurements are possible.

APPENDIX II

2 MEASUREMENTS OF SUSPENDED LOAD

2.1.1 General

Glacier streams transport great loads of sediment (silt, sand, gravel, rocks and boulders) that results from glacier erosion. It is mainly carried in suspension, and some is moved along the bed.

The bedload is only a relatively small part of the total material transported in the stream, good information about the rate of glacier erosion can be obtained by taking samples of river water, analyzing it for suspended material, and calculating the sediment transport. The bedload transport of boulders, is difficult to observe and, until reliable methods for its determination have been developed, can not be estimated. It is assumed that the suspended load accounts for about 40 to 50 % of the total load carried by the glacier stream-at the ice front. The amount of suspended material is relatively easy to determine, as it can be calculated from known water discharge and laboratory analysis of water samples.

Results of sediment studies will not only give valuable information about glacier erosion under various conditions but also indicate the possible rates of sedimentation that can be expected in reservoirs and lakes along rivers from glacierized areas. Natural or artificial lakes are greatly affected by sedimentation and, in the case of water utilization, the capacity of storage dams will be gradually reduced (e.g.Ostrem and Olsen, 1987). The smaller fractions which pass through the lakes and reservoir will, in the case of water-power stations, drastically increase the wear of turbines, etc. Thus, studies of sediment transport in glacier streams are of direct practical importance.

2.1.2 Location of Sampling Site

Water samples should be taken close to the glacier snout. Sampling site preferably selected anywhere the flow is turbulent so that the sample is representative of the total discharge at the time of sampling. This means that the sample could be taken just below a small waterfall or in a section of extremely turbulent water.

2.1.3 Sampling Method

Automatic water samples may be used to collect meltwater samples. The sampler can be programmed to collect water sample every hour. The manual collection of water samples in the Himalaya is very difficult because of logistical problems. Figure 2A.1 shows auto-sampler in op-

eration on Bhagirathi river close to the Gangotri glacier snout, Garhwal Himalaya. The samples should be immediately filtered in the field after collection and the quantity of sediment retained should be determined. Suspended sediment flux (kg s^{-1}) could be obtained as the product of sampled sediment concentration (gl^{-1}) and hourly average discharge (m^3 s^{-1}). Daily total sediment load could be computed from the 24 h suspended sediment flux values. For calculation of total sediment yield in a summer, values of total daily load on which 24 h sediment load was measured should be used.

For solute concentrations studies meltwater samples should be immediately vacuum filtered in the field through a 0.45 m filter membrane, using hand pump. Three 5 ml aliquot of sample should be used to clean the filter paper and apparatus. An aliquot of 150 ml should be rapidly filtered and decanted into two polypropylene bottles, which were rinsed three with filtrate. The first rapidly filtered solution should be used for pH and alkalinity determinations. A second 150 ml aliquot of sample filtered and decanted into two rinsed bottles. One solution should be acidified with 1 ml of 50% HNO_3 and should be used for cation analysis, while the other unacidified solution should be used for the determination of anions. The samples should be stored at about 4°C in the cold room during their analysis at the laboratory.

2.1.4 Numbering of Samples

A special numbering system has been developed for sediment samples. Each number consists of eight digits, the first four indicate the data, the last four the time of the day when the samples was taken. *Example*: 07131145 means that the sample was taken on July 13 at 11.45 hrs.

2.2 ELECTRICAL CONDUCTIVITY

2.2.1 Use in water chemistry studies

Electrical conductivity is often used in water quality investigations as a surrogate measure of the total dissolved solids (TDS) content of water. It provides a valuable indication of total ionic concentration, and is useful in the monitoring of temporal and spatial variations of TDS at a station and throughout a catchment respectively. It is only an indicator of TDS and id perhaps best utilized as comparative parameter, e.g. to show differences in the quality of waters draining from different subglacial pathways. Electrical conductivity values provide a useful check on other analytical data for specific ions.

Continuous monitoring conductivity, as a surrogate for total dissolved solids, has provided a simple means of separating hydrographs into quick and delayed flow components so that inferences could be made about the routing and residence times of those flow components within the glacier basin.

BIBLIOGRAPHY

Addison, K. 1981. The contribution of discontinous rock-mass failure to glacier erosion, *Ann. of Glaciol.,* **2,** 3-10.

Agassiz, J.L.R. 1840. *Etudes "Sur les Glace",* (privately published), Neuchantel, Switzerland, pp.161.

Ageta, Y. and Higuchi, K., 1984. Estimation of mass balance components of a summer-accumulation type glacier in the Nepal Himalaya. *Geografis. Annal.,* **66 A,** 249-255.

Ageta, Y., 1983. Characteristics of mass balance of summer-accumulation type glacier in the Nepal Himalaya (1), (11). *Seppyo,* **45,** 81-105.

Ageta, Y., and Pokhrel, P. 1999. Glacier research in the Nepal Himalaya. *Moun. Res. Deve. Special Volume on Himalayan Glaciology,* Hasnain, S.I. (Ed.) California Press, USA.

Ageta, Y., Ohata, T., Tanaka, Y., Ikegami, K. and Higuchi, K. 1980. Mass balance of glacier AX010 in Shorong Himal, east Nepal during the summer monsoon season. *Seppyo, Special issue,* **41,** 34-41.

Alley, R.B. Blankenship, D.D., Bentley, C.R., and Rooney, S.T. 1987a. Till beneath ice stream B.3. Till deformation: evidence and implications. *J. Geophys. Res.,* **92** (B9), 8921-8929.

Alley, R.B., Blakenship, D.D., Rooney, S.T. and Bentley, C.R. (1987) Deformation of till beneath ice stream B, West Antartica. *Nature,* **322,.** 57-59.

Alley, R.B., Blakenship, D.D., Rooney, S.T. and Bentley, C.R. 1987b. Till beneath Ice stream B.4. A coupled ice-till flow model. *J. Geophys. Res.,* **92** (B9), 8921-8929.

Andrews, J.T. 1972 Glacier power, mass balances, velocities and erosion potential. *Z. Geomorphol.,* **13,** 1-17.

Andrews, J.T. 1992. Sedimentary geofluxes, global. In Nierenberg, W.A. (Ed.), Encyclopedia of Earth System Science, **4.** San Diego: Academic Press, 93-101.

Atkinson, E.T. 1981. The *Himalayan Gazetteer,* **111,** Parts 1 and 11, Cosmo Publications, New-Delhi.

Auden, J.B. 1935. The snout of the Biafor Glacier in Bultistan. *Rec. Geol. Surv. of India.* **38,** 400-13.

Auden, J.B. 1937. The snout of the Gangotri glacier, Tehri Garhwal. *Rec. Geol. Surv. of India.* **72,** Pt 2, 125-140.

Bandyopadhyay, J. and Gyawali, D. 1994. Himalayan water Resources: ecological and political aspects of management, *Mountain Research and Development,* **14. (No.1),.** 1-24.

Barrie, L.A. 1991. Snow formation and processes in the atmosphere that influence its chemical composition. In (Davies, T.D. Tranter. M. and Jones, H.G., (Eds) Seasonal Snowpacks: Processes of Compositional Change. Springer-Verlag, Berlin, 1-20.

Barry, R. 1992. Mountain Weather and Climate, Second, Edition, Routledge, London.

Batten, J.H. 1845. 'On the Line of Perpetual Snow in the Himalaya'. Calcutta *J. Nat. Hist.* **5,** 383-8.

Beecroft, I. 1983. Sediment transport during an outburst from Glacier de Tsidjiore Nouve, Switzerland, 16-19 June 1981. *J. Glaciol.,* **29,** (101), 185-190.

Behrens, H., Bergman, H., Moser, H., Rauert, W., Stichler, W. and Ambach, W. 1971. Study of the discharge of Alpine glaciers by means of environmental isotopes and dye tracers. *Z. Glets. and Glazialgeolo.,* **7 (1, 2),** 79-102.

Bezinge, A. Clark, M.J., Gurnell, A.M. and Warburton, J. 1989. The management of sediment transport ported by glacial meltwater streams and its significance for the estimation of sediment yield. *Ann. Glaciol.,* **13,** 15.

Bindschadler, R. 1983. The importance of pressurized subglacial water in separation and sliding at the glacier bed. *J. Glaciol.,* **29(101),** 3-19.

Blanford, W.T. 1877. Remarks on Himalayan glacier. *Proc. Asia. Soc. Beng.,* 3-4.

Blanford, W.T. 1891. Notes on age and ancient glaciers of Himalaya. *Geological Magazine,* **209 (10,8)**, 372-375.

Blanford, W.T. 1873. Notes on glaciers in Hindustan. *Nature,* **9,** 63.

Bogen, J. 1980. The hysteresis effect of sediment transport systems. *Nor. Geogr. Tidsskr.,* **34(1),** 45-54.

Bogen, J. 1991. Erosion and sediment transport in Svalbard. In Gjessing, Y; Hagen, J.O; Hassel, K.S; and Wold, B, Eds. Arctic Hydrology: present and future tasks. Hydrology of Savlbard-hydrological problems in a cold climate, Oslo, *Norwegian National Committee of Hydrology,* **23,** 147-158.

Bogen, J. 1996. Erosion rates and sediment yields of glaciers. *Ann. Glaciol.,* **2,** 48-52.

Borland, W.M. 1961. Sediment transport of glacier-fed streams in Alaska. *J. Geophys. Res.,* **66,** 3347-3350.

Boulton, G.S. 1975. Processes and patterns of subglacial sedimentation: A theortical approach. In: Wright, A.E. and Moseley, F. (Eds.) Ice Ages Ancient and Modern, *Geological Journal special issue,* **6,** Seel House Press, Liverpool, 7-42.

Boulton, G.S. 1982. Subglacial processes and the development of glacial bedforms. In: *6th Guelph Symposium on Geomorphology,* Davidson-Arnott, R., Nickling, W. and Fahey, B.D. (Eds.). Research in glacial, glacio-fluvial and glacio-lacustrine systems, Norwich, pp. 1-31.

Boulton, G.S. 1987. A theory of drumlin formation by subglacial sediment deformation. In: *Drumlin Symposium,* Menzies, J and Rose, J. (Eds.). A.A. Balkema, Rotterdam, pp. 25-80.

Boulton, G.S. 1974. Processes and patterns of glacial erosion. In: *Glacial Geomorphology,* D.R. Coates (Ed.). New York State university, New York, 41-87.

Boulton, G.S. 1978. Boulder Shape and grain-size distributions of debris as indicators of transport paths through a glacier and till genesis. *Sedimentology,* **25,** 773-799.

Boulton, G.S. 1979. Processes of glacial erosion on different substrata. *J. Glaciol.,* **22(88),** 15-38.

Boulton, G.S. 1996. Theory of glacial erosion, transport and deposition as a consequence of subglacial deformation. *J. Glaciol.,* **42,** 43-62.

Boulton, G.S., Morris, E.M., Armstrong, A.A. and Thomas, A. 1979. Direct measurement of stress at the base of a glacier. *J. Glaciol.,* **22(86),** 3-24.

Braithwaite, R.J. and Olesen, O.B. 1990. A simple energy-balance model to calculate ice ablation the margin of the Geenland ice sheet, *J. Glaciol.,* **36(123),** 222-228.

Brown, G.H. 1991. Solute provenance and transport pathways in Alpine glaciers. (Ph.D. thesis, University of Southampton, U.K.).

Brown, G.H. and Tranter, M. 1990. Hydrograph and chemograph separation of bulk meltwaters draining the Upper Arolla glacier, Valais, Switzerland. Proc. Symp. *Hydrology in Mountain Region. I. Hydrological Measurements; the WaterCycle Int. Asso. Hydrol. Sci. Pub.* **(193, No. 6).** 429-437.

Brown, G.H. Sharp, M.J. Tranter, M. Gurnell, A.M. and Nienow, P.W. 1994. Impact of post-mixing chemical reactions on the major ion chemistry of bulk meltwaters draining the Haut Glacier d'Arolla, Valais, Switzerland. *Hydrolog. Processes,* **8,** 465-480.

Brown, G.H., Sharp, M. and Tranter, M. 1996. Subglacial chemical erosion: seasonal variations in solute provenance, Haut Glacier d'Arolla,

Valais, Switzerland. *Ann. Glaciol.* **22,** 25-31 (Proceedings of the International Symposium on Glacial Erosion and Sedimentation, held in Reykjavik, Iceland, 20-25 August 1995).

Bruijnzeel, L.A. and Bremmer, C.N. 1989. High level-Low level inter activities in the Ganges-Brahmaputra river basin: A review of published literature. *ICIMOD Occasional paper No. 11,* Kathmandu, Nepal.

Brunsden, D. 1979. Weathering. In: *Processes in Geomorphology,* Embleton, C. and Thornes, J.B. (Eds.), Arnold, London, pp. 73-129.

Budd, W.F. and Allison, I.F. 1975. An empirical scheme for estimating the dynamics of unmeasured glaciers. In: *Snow and Ice* (Proc. Moscow Symp., August 1971), International Association of hydrological Sciences, Publ. **No. 104,** 246-256.

Budd. W.F. and Jensson, D. 1975. Numerical modeling of glacier system. In: Snow and Ice (Proc. Moscow, Symp., August 1971), 257-291. *Int. Asso. Hydrlo. Sci. Pub.* **No. 104,** 16-32.

Busenberg, E. and Clemency, C.V. 1976. The dissolution kinetic of feldspars at 25°C and 1 atms CO_2 partial pressure. *Geochim. Cosmochim. Acta,* **40,** 41-49.

Cadle, S.H. 1991. Dry deposition to snowpacks. In: *Seasonal Snowpacks: Processes of Compositional Change,* Davies, T.D., Tranter, M. and Jones, H.G. (Eds), Springer-Verlag, Berlin, 21-65.

Carline, R.F., Dc Walle, D.R., Sharpe, W.E., Dempsey, B.A., Gagen, C.J. and Swistock, B. 1992. Water chemistry and Fish community responses to episodic stream acidification in Pennysylvania, USA, *Environment pollution,* **78,** 45-48.

Carson, M.A. 1971. *The Mechanics of Erosion*-Momographs in spatial and environmental system analysis, Pion, London, pp. 174.

Casey, W.H. and B. Bunker. 1990. Leaching of mineral and glass surfaces during dissolution. *Reviews in Mineralogy,* **23,** 397-426.

Chalise, S.R. 1994. Mountain environments and Climate Change in the Hindu Kush-Himalaya. In: *Mountain Environments in changing climate,* Martin Beniston, (Ed.), Routledge, pp. 382-404.

Chamberlin, T.C. 1878. The extent and significance of the Wisconsin Kettle moraines. *Wisconsin Acad. Sci. Trans.,* **4,** 201-234.

Chauhan, D.S. and Hasnain, S.I. 1993. Chemical characteristics, solute and suspended sediments load in the meltwaters draining Satopanth and Bhagirathi khark glaciers, Ganga basin, India. *International Association of Hydrological Sciences, Pub.* **No. 218,** 403-410.

Chou, L. and R. Wollast. 1984. Study of weathering of albite at room temperature and pressure with a fluidized bed reactor. *Geochim. Cosmochim. Acta,* **48,** 2205-2217.

Church, M. 1974. On the quality of some waters on Baffin Island, Northwest Territories. *Chand. J. Earth Sci.,* **11, No. 12,** 1676-1688.

Church, M. and Ryder, J.M. 1972. Paraglacial sedimantation: a consideration of fluvial processes conditioned by glaciation. *Geolog. Soc. Ameri. Bull.,* **83,** 3059-3071.

Church, M. and Slaymaker, O. 1989. Disequilibrium of Holocene sediment yield in glaciated British Columbia. *Nature,* **337,** 452-454.

Clague, J.J. 1986. The quaternary stratigraphic record of British Columbia-evidence for episodic sedimentation and erosion controlled by glaciation. *Cana. J. Earth Sci.,* **23,** 885-894.

Clayton, L. and Moran, S.R. 1974. A glacial process form model. In: *Glacial Geomorphology,* Coates, D.R. (Ed.), State University of New York, Binghampton, 88-119.

Collins, D.N. 1978. Hydrology of an Alpine glacier as indicated by the chemical composition of meltwater. *Z. Gletscherked. Glazialg.,* **13(1/2),** 219-238.

Collins, D.N. 1979a. Hydrochemistry of Meltwaters draining from an Alpine glacier. *Arct. & Alpine Rese.,* **11, (No. 3),** 307-324.

Collins, D.N. 1979b. Quantitative determination of the subglacial hydrology of two Alpine glaciers. *J. Glaciol.,* **23, (No. 89),** 347-362.

Collins, D.N. 1979c. Sediment concentration in meltwater as an indicator of erosion processes beneath an, Alpine glacier. *J. Glaciol.,* **23, (No. 89),** 247-257.

Collins, D.N. 1981. Seasonal variation of solute concentration in meltwaters draining from an Alpine glacier. *Ann. Glaciol.*, **2,** 11-16.

Collins, D.N. 1982. Water storage in an alpine glacier, In: *Hydrological Aspects of Alpine and High Mountain Areas.* Proceedings of the Exeter Symposium, Glen, J.W. (Ed.), International Association of Hydrological Sciences, Pub. **No. 138,** 113-122.

Collins, D.N. 1983. Solute yield from glacierised high mountain basin. International Association of Hydrological Sciences, Pub. **No. 141,** 41-50.

Collins, D.N. 1991. Climatic and glaciological influence on suspended sediment transport from an Alpine glacier, In: *sediment and stream water quality in a changing environment: trends and explanation* (Proceedings Vienna Symposium., 1991), International Association of Hydrological Sciences, Pub. **No. 202**-3-12.

Collins, D.N. 1995a. A conceptual-based model of subglacial sediment entrainment by flowing water. *Working paper 19, University of Manchester, U.K.*, pp. 15.

Collins, D.N. 1995b. Dissolution Kinetics Transit times through subglacial Hydrological pathways and Diurnal Variations of solute content of Meltwaters Draining from an Alpine glacier. *Hydrolog. Processes,* **9,** 897-910.

Collins, D.N. 1996. Sediment transport from glacierised basins in the Karakoram Mountains. *Working paper 20, University of Manchester, U.K.*, pp. 18.

Collins, D.N. and Hasnain, S.I. 1994. Runoff and sediment transport from glacierised basins at the Himalayan scale. *Working paper 15, University of Manchester, U.K.*, pp. 14.

Collins, D.N. and Hasnain, S.I. 1995. Runoff and sediment transport from the glacierised basins at the Himalayan scale. In: *Effects of Scale on Interpretation and Management of Sediment and Water Quality,* International Association of Hydrological Sciences, Pub. **No. 226**, 17-25.

Collins, D.N. and Young, G.J. 1981. Meltwater hydrology and hydrochemistry in snow and ice-covered mountain catchments. *Nord. hydrology,* **12 (No. 4-5)**, 319-334.

Cunninham, J.W. 1849. Notes on the Limits of Prepetual Snow in the Himalaya. J. Asi. *soc. Bengal,* **18,** 694-697.

Davies, T.D., Vincent, C.E. and Brimblecombe, B. 1982. Preferential elution of strong acids from a Norwegian ice cap. *Nature, London,* **300,** 161-163.

Desio, A. 1954. An Exceptional Glacier Advance in the Karakoram-Ladakh region. *J. Glaciol.* **2 (No. 15)**, 383-386.

Drewry, D. 1986. Glacial geologic processes. London, Edward Arnold, pp. 276.

Dutt, G.N. 1961. The Bara-Shigri Glacier, Kangra District, East Punjab, India. *J. Glaciol.,* **3,** 1007-1015.

Egerton, P.H. 1864. *Journal of a Tour through Spiti,* Cundeall Downes, London.

Elliston, G.R. 1973. Water movement through the Gornergletscher. In: *Symposium on the Hydrology of Glaciers, Cambridge, 1969.* International Association of Hydrological Sciences, Pub. **No. 95,** 79-94.

Embleton, C. and King, C.A.M. 1968. Glacial and Periglacial Geomorphology. Edward Arnold, London, pp. 608.

Embleton, C. and King, A.M. 1975. Glacial Geomorophology. Edward Arnold, London, pp. 537.

Everest, R. 1832. Some observations on the quantity of earthy matter brought down by the Ganga river. *Royal Asiatic Society, Bengal.*

Eyles, N. Sasserville, D.R. Slatt, R.M. and Rogerson, R.J. 1982. Geochemical denudation rates and solute transport mechanism in a maritime temperate glacier basin. *Cana. J. Earth Sci.,* **19,** 1570-1581.

Feininger, T. 1971. Chemical weathering and glacial erosion of crystalline rocks and the origin of till. *United States Geolo. Surv. Professional Paper,* **750c,** C65-C81.

Fenn, C.R. 1987. Sediment Transfer Processes in Alpine Glacier Basins. In: *Glacio-Fluvial Sediment transfer-An Alpine Perspective.* Gurnell, A.M. and Clark, M.J. (Eds.) John Wiley and Sons, Chichester, 59-85.

Fenn, C.R. 1989. Quantifying errors involved in transferring suspended sediment rating equations across ablation seasons. *Ann. Glaciol.,* **13,** 64-68.

Ferguson, R.I. 1984. Sediment load of the Hunza River. In: *International Karakoram Project 2.* Miller, K.J. (Ed.), Cambridge University Press, Cambridge, 581-598.

Ferguson, R.I. 1986. River loads understimated by rating curves. *Water Reso. Res.* **22,** 1: 74-76.

Forbes, J. 1842. Account of his recent observations on glaciers. *New Philosophical journal,* **33,** 38-352.

Ford, D.C. and Williams, P. 1989. Karst Geomorphology and Hydrology. Unwin Hyman, London, pp. 601.

Fountain, A.G. 1992. Subglacial water flow inferred from stream measurements at South Cascade Glacier, Washington, U.S.A. *J. Glaciol.,* **38,** 51-64.

Fowler, A.C. 1987. Sliding with cavity formation. *J. Glaciol.,* **33 (No. 115),** 255-267.

Fujii, Y. 1977. Glaciers (in Japanese). In: Kawakita. J. (Ed), *The Himalaya,* Asahi Shinbunsya, Tokyo, 136-144.

Fujii, Y., and Higuchi, K. 1977. Statistical analyses of the forms of the glaciers in the Khumbu Himal. *Seppyo, special issue,* **39,** 7-14.

Fukushima, Y., Kawashima, K., Suzuki, M., Ohta, T., Motoyama, H., Kubota, H., and Bajracharya, O.M. 1987. The hydrological data of Langtang valley, Nepal Himalaya. *Bull. Glacier Res.* **5,** 115-120.

Fushimi, H., Yoshida, M., Watanabe, O. and Upadhyay, B.P. 1980. Distribution and grain sizes of supraglacial debris in the khumbu glacier, Khumbu region, East Nepal. *Seppyo special issue,* **41,** 18-25.

Gardner, J.S. 1986. Recent fluctuations of Rakhoit glacier, Nanga Parbat, Punjab Himalaya, Pakistan. *J. Glaciol.,* **33 (No. 112),** 527-529.

Gardner, J.S., and Jones, N.K. 1990. Sediment transport and yield at the Rakhiot glacier, Nanga Parbat, Punjab Himalaya. In: *Snow and Ice Hydrology Project working Annual Report and Scientific papers. Wilfrid Laurier University,* pp. 179.

Garrels, R.M. and Christ, C.L. 1965. Solutions, Minerals and Equilibria. Freeman Cooper, San Francisco, pp. 234.

Garrels, R.M. and Howard, P. 1956. Reactions of feldspar and mica with water at low temperature and pressure. *Clays & Clay Miner.,* **6,** 68-88.

Gilbert, G.K. 1903. Glaciers and Glaciations of Alaska, *Harriman Alaska Series,* 3. Doubleday Page & Company, New York. pp. 231.

Gilbert, G.K. 1906. Crecentic gouges on glaciated surfaces. *Geol. Soc. America Bull.,* **17,** 303-316.

Glen, J.W. 1954. The stability of ice-dammed lakes and other water-filled holes in glaciers. *J. Glaciol.,* **2,** 316-318.

Glen, J.W., and Lewwis, W.V. 1961. Measurement ofs of side-slip at Austerdalsbreen, *J. Glaciol.,* **3,** 1109-1112.

Godwin - Austin, H.H. 1862. On the Glaciers of the Mustakh Ranges'. *Jour. R. Geogr. Soc.* 34: 19-56.

Gornitz, V., Lebedeff, S. and Hansen, J. 1982. Global sea level trend in the past century. *Science,* **215,** 1611-1614.

Griesbach, 1891. Geological traverse in the Gangotri region. *Mem. Geol. Surv. India,* **XXIII**, 27.

Griesbach, C.L. 1887. Recent Glacial Action in the Hindu Kush. *Nature,* **35,** 594.

Gulati, T.D. 1973. Role of snow and ice hydrology in India. *Proc. of Int. Symp. on the Role of Snow and Ice Hydrology, Banaiff, Sept. 1972, Int. Glaciol. Soc.,* 610-623.

Gunn, J.P. 1930. Report on the Khumdan Dam and Shyok Flood of 1929. Govt. Punj. Publ., Lahore.

Gurnell, A.M. 1982. The Dynamics of suspended sediment concentration in a proglacial stream. In: Hydrological Aspects of Alpine and High Mountain Areas (Proceedings of the Exeter Symposium, July 1982) Glen J.W. (Ed.). *International Association of Hydrological Sciences, Pub.* **No. 138,** 319-330.

Gurnell, A.M. 1987. Suspended sediments. In: *Glacio-Fluvial Sediment Transfer: An Alpine Perspective,* Gurnell, A.M. and Clark, M.J. (Eds), John Wiley and Sons, Chichester, 305-354.

Gurnell, A.M., Brown, G.H. & Tranter, M. 1994. Sampling strategy to describe the temporal hydrochemical characteristics of an Alpine proglacial stream. *Hydrolog. Process,* **8,** 1-25.

Gurnell, A.M. and Fenn, C.R. 1984. Flow spearation sediment source areas and suspended sediment transport in a proglacial stream, *catena supplement* **5,** 109-119.

Gurnell, A.M. 1984b Fenn, C.R. 1984. Box-Jenkins transfer function models applied to suspended sediment concentration-discharge relationships in a proglacial stream. *Arct. & Alp. Res.,* **16,** 93-106.

Gurnell, A.M. and Fenn, C.R. 1985. Spatial and temporal variations in electrical conductivity in a proglacial stream system. *J. Glaciol.,* **31,** 108-114.

Gurnell, A.M., Hodson, A., Clark, M.J., Bogen, J., Hagen, J.O. and Tranter, M. 1994. Water and sediment discharge from glacier basins: and Arctic and Alpine comparison. International Association of Hydrological Sciences, Pub. No. 224-299-311.

Gurnell, A.M. and Warburton, J. 1990. The significance of suspended sediment pulse for estimating suspended sediment load sediment and identifying suspended sediment sources in alpine glacier basins. Proc. Inter. Conference on Water Resources in Mountainous Regions, *International Association of Hydrological Sciences, Pub.* **No. 193,** 463-470.

Gurnell, A.M. Warburton, J. & Clark, M.J. 1988. A comparison of the sediment transport and yield characteristics of two adjacent glacier basins, Val d'Heren Switzerland. *International Association of Hydrological Sciences, Pub.* **No. 174,** 431-441.

Guymon, G.L. 1974. Regional sediment yields analysis of Alaska streams. *American Soc. of Civil Engineers, J. Hydra. Division,* **100,** 41-51.

Haeberli, W. 1990. Glacier and permafrost signals of 20th century warming. *Ann. Glaciol.,* **14,** 99-101.

Haeberli, W., Hoelzle, M. & Suter, S. (Eds.). 1998. Into the second century of worldwide glacier monitoring: prospects and strategies. *Studies and Reports in Hydrology* **56,** *pp. 257, UNESCO,* Paris, France.

Hagen, J.O., Wold, B., Wiestol, O., Ostrem, G. and Sollid, J.L. 1983. Subglacial processes at Bandhusbreen, Norway: preliminary results, *Ann. Glaciol.,* **4,** 91-98.

Hall, F.R. 1970. Dissolved solids-discharge relationships, 1, Mixing models. *Water Resour. Res.,* **6,** 845-850.

Hallet, B. 1976. Deposits formed by subglacial precipitation of CaCO3. *Geol., Soci., Am. Bull.,* **85,** 1003-1015.

Hallet, B., Hunter, L.E. and Bogen, J. 1996. Rates of erosion and sediment evacuation by glaciers: a review of evidence. *Global & Planet. Change,* **12,** 213-235.

Hallet, B. *et. al.,* 1978. The composition of basal ice from a glacier sliding over limestones. *Geol. Soc. Am. Bull.,* **89,** 314-320.

Hammer, K.M. and Smith, N.D. 1983. Sediment production and transport in a proglacial stream, Hilda creek, Alberta, Canada. *Boreas,* **12,** 91-106.

Hantz, D. and Lliboutry, L. 1983. Waterways, ice permeability, and water pressures at glacier d'Argentiere, French Alps, *J., Glaciol.,* **29 (No. 102)**, 227-239.

Harbor, J.M. 1990. Glacial-valley cross-section development under conditions of temporally varying ice discharge: results of a numerical model. *Geol. Soci. Am. Abstracts,* **22** (No. 7), A108.

Harbor, J.M. 1992. Numerical modelling of the development of U-shaped valleys by glacial erosion. *Geol. Soci. Am. Bull.,* **104,** 1364-1375.

Harbor, J.M. and Warburton, J. 1992. Glaciation and denudatiion rates. *Nature,* 365-751.

Harbor, J.M. and Warburton, J. 1993. Relative rates of Glacial and nonglacial erosion in Alpine Environments. *Arct. & Alpine Res.,* **25 (No. 1)**.

Hasnain, S.I. 1989. Himalayan Glaciers as a sustainable water resource. *Water Resour. Deve.,* **5 (No. 2)**, 106-112.

Hasnain, S.I. and Chauhan, D.S. 1993. Sediment transfer in the glaciofluvial environment-A Himalayan perspective. *Environ. Geol.,* **22**(3), 205-211.

Hasnain, S.I. and Thayyen, R.J. 1999a. Controls on the major-ion chemistry of Dokriani glacier meltwaters, Ganga basin, Garhwal Himalaya, India. *J. Glaciol.,* **45**, No. 149, pp. 87-92.

Hasnain, S.I. and Thayyen, R.J. 1999b. Discharge and suspended-sediment concentration of meltwaters, draining from the Dokriani glacier, Garhwal Himalaya, India. J. Hydrology, 218, pp. 191-198.

Hasnain, S.I. Subramanian, V. and Dhanpal, K. 1989. Chemical characteristics and suspended sediment load of meltwaters from a Himalayan glacier in India. *J. Hyd.,* **106,** 99-106.

Hasnain, S.I. and Thayyen, R.J. 1994. Hydrograph separation of bulk meltwaters of Dokriani Banak Glacier basin, based on electrical conductivity, *Current Science,* **67 (No. 3)**, 189-193.

Hasnain, S.I. and Thayyen R.J. 1996. Variation of Discharge and Solute concentration in the Meltwaters of Dokriani Glacier, Garhwal Himalaya, India. *J. Geol. Soci. India.,* **47,** 89-93.

Hasnain, S.I. 1996. Factors controlling suspended sediment transport in Himalayan glacier meltwaters. *J. Hydrol.,* **181,** 49-62.

Helland, A. 1877. On the ice-fjords of North-Greenland, and on the formation of fjords, lakes and cirques in Norway and Greenland. *Quart. J. Geol. Soc. London,* **33,** 142-176.

Hem. J.D. 1970. Study and interpretation of the chemical characteristics of natural water. *U.S. Geol. Surv. Water Supply Pap.*, 1473(2nd edi). pp. 269.

Hewitt, Kenneth, 1969. Glacier Surges in the Karakoram Himalaya. *Can. J. Earth Sci.*, **6 (No. 4),** pt. 2, 1009-1018.

Hewitt, K. 1982. Natural dams and outburst floods of the Karakoram Himalaya. In: Hydrological Aspects of Alpine and High Mountain Areas. Glen J.W. (Ed.), *International Association of Hydrological Sciences, Pub. (No. 138),* 259-269.

Hicks, D.M. McSaveney, M.J. and Chinn, T.J.H. 1990. Sedimentation in proglacial Ivory Lake, Southern Alps, New Zealand. *Arct. & Alpine Res.,* **22 (No. 1),** 26-42.

Higuchi, K. Ageta, Y. Yasunari, T. and Inone, J. 1982. Characteristics of precipitation during the monsoon season in high-mountain areas of the Nepal Himalaya. In: Hydrological Aspects of Alpine and High Mountain Areas (Proceedings of the Exeter Symposium, July 1982). *International Association of Hydrological Sciences, Publ. No. 138,* 21-30.

Higuchi, K., Ageta, Y. and Kodama, H. 1976. Water discharge of Imja Khola in Khumbu Himalaya. *Seppyo, Special Issue,* **38,** 22-26.

Higuchi, K., Fushimi, H., Ohata, T., Iwata, S., Yokoyama, K., Higuchi, H., Nagoshi, A. and Iozawa, T. 1980. Glacier inventory in the Dud Kosi region, East Nepal. In the World Glacier Inventory. *International Association of Hydrological Sciences, Publ. No. 218,* 29-36.

Hochella, M.F. and A.F. White. 1990. Mineral-water interface geochemistry. *Revi. in Mineralogy,* **23,** 5-10.

Hodgkins, R. 196. Seasonal trend in suspended-sediment transport from an Arctic glacier, and implications for drainage-system structure. *Ann. Glaciol.,* **22,** 147-151.

Hoinkes, H. 1976. Glacial meteorology. In: *Research in Geophysics,* Odishaw, H. (Ed.), **2,** MIT Press, Cambridge, Messachusetts, 391-424.

Holland, H.D. 1978. The chemistry of atmospheres and oceans. New York, Wiley-Interscience. pp. 324.

Hubbard, B. and Nienow, P. 1997. Alpine subglacial hydrology. *Quatern. Sci. Revi.,* **16.** 939-955.

Humbolt, A. 1845. Remarks on the height of the snowline in the Himalaya'. *Cosmos,***1,** 358, 483-4.

IAHS (ICSI)/UNEP/UNESCO. 1993. *Fluctuations of Glaciers* 1985-90. Haeberli, W. and Hoelzle, M. (eds.). *UNESCO, Paris.*

Iken, A. 1981. The effect of the subglacial water pressure on the sliding velocity of a glacier in an idealized numerical model. *J. Glaciol.,* **27(No. 97)**, 407-421.

Iken, A., Rothlisberger, H., Flotron A. and Haeberli, W. 1983. The uplift of Unteraargletscher at the begining of the melt season - a consequence of water storage at the bed? *J. Glaciol.,* **29 (No. 101)**, 28-47.

Inoue, J. 1977. Mass budget of Khumbu glacier. *Seppyo, Special issue,* **39,** 15-19.

Inoue, J., and Yoshida, M. 1980. Ablation and heat exchange over the Khumbu glacier, *Seppyo, special issue,* **41,** 26-33.

Iverson, N.R. 1991. Potential effects of subglacial water-pressure fluctuations on quarrying. *J. Glaciol.,* **37**(No. 125). 25-36.

Ives, J.D. and Messerli, B. 1989. The Himalayan Dilemma: Reconciling Development and Conservation. United Nations University and Routledge, London, pp.

Ives, J.D. 1973. Arctic and Alpine geomorphology - a review of current outlook and notable gaps in knowledge. In: *Research in Polar and Alpine Geomorphology,* 3rd Guelph Symposium on Geomorphology, Fahey, B.D. and Thompson, R.D. (Eds) Geo Books, Norwich, 1-10.

Jangpangi, B.S. 1958. Study of some of the central Himalayan glaciers. *J. Sci. Ind. Res.* **17 (No. 12),** 91-3.

Jansson, M. 1988. A global survey of sediment yield. *Geografis. Ann.*, **70A,** 81-98.

Johannessen, M. and Henriksen, A. 1978. Chemistry of snow meltwater: changes in concentration during melting. *Water Resour. Res.*, **14**(4), 615-619.

Johannesson, T., Sigurosson, O., Laumann, T., and Kennett, M. 1993. Degree-day glacier mass balance modelling with applications to glaciers in Iceland and Norway. *NORDIC Hydrol.Progr. NHP Report No. 33,* pp. 40.

Jones, H.G., Tranter, M. and Davies, T.D. 1989. The leaching of strong acid anions from snow during rain-on-snow events: evedence for two component mixing. *International Association of Hydrological Sciences, Pub. 179* (Symposium at Baltimore 1987 — Atmospheric Deposition), 239-250.

Jopling, A.V. and McDonald, B.C. Glaciofluvial and Glaciolacustrine sedimentation. Tulsa, Okla: Society of Economic Palaeontologists and Mineralogists, special publication **23**, pp. 101-112.

Jordan, P. and Slaymaker, O. 1991. Holocene sediment production in Lilloef River basin, British Columbia: A sediment budget approach. *Geogr. Physi. et. Quatern.*, **45,** 45-47.

Kadota, T., Seko, K., and Ageta, Y. 1993. Shrinkage of glacier AX 010 since 1978, Shorong Himal, East Nepal. *International Association of Hydrological Sciences, Pub.* **No. 218,** 145-154.

Kanwar Sain, 1946. The Role of Glaciers and Hydrology of Punjab Rivers, *Central Board of Irrigation, Pub.* **No. 36,** New Delhi, pp. 29.

Kasser, P. 1959. Der Einfluss von Gletscherruckgang und Gletschervostoss auf den wasserhaushalt, *Wasser und Energiewirt.*, **6,** 155-168.

Keller, W.D. and Reesman, A.L. 1963. Glacial milks and their laboratory simulated counterparts. *Geol. Soc. of Am. Bull.*, **74,** 61-76.

Kick, W. 1989. The decline of the last little Ice Age in High Asia compared with that in the Alp, in: *Glacier fluctuations and climatic change,*

Oerlemans. J. (Ed.), 129-142, Kluwer Academic Publishers, Dordrecht.

Knudsen, N.T. 1990. Dissolved load in glacier river water — Austre okstindbreen, *Okstindan Norway report no 90.2,* Unpublished.

Kotlyakov, V.M. 1996. Variations of snow and Ice in the past and at present on a global and regional scale. IHP-V. *Technical Documents in Hydrology No. 1* UNESCO, Paris. Krenek, L and Bhawani, V. 1945. Recent and Past glaciation of Lahul, Indian. *Geog. J.* **20(3)**, 93-102.

Kuhn, M. 1985. Reactions of mid-latitude glacier mass balance to predicted climatic changes, In: Glaciers, Ice Sheets and Sea level: Effect of CO_2 Induced Climatic Change. Report of a Workshop in Seattle, 1984, *National Technical information Service, U.S. Department of Commerce Springfield, VA,* 248-254.

Kukla, G and Gavin, J. 1981. Summer Ice and carbon dioxide. *Science,* **214,** 497-503.

Lang, H. 1973. Variations in the relation between glacier discharge and meteorological elements. In: Symposium on the Hydrology of Glaciers (Proceedings of the Cambridge Symposium, 7-13 September 1969), *International Association of Scientific Hydrology Pub.* **No. 95,** 85-94.

Lang, H. 1980. Theortical and practical aspects in the computation of runoff from glacier areas. In: Proceedings of the Symposium in Tblisi 1978. *Geophysical Committee, data of Glaciological studies, USSR Academy of Sciences, Publication 38,* 187-194.

Lang, H., Leibundgut, C., and Festel, E. 1979. Results from tracer experiments on the water flow through the Aletschgletscher, *Z. Gletscherk. Gletscher. & Glazialg.,* **15,** 209-218.

Lang, H., Schadler, B., and Davidson, G. 1977. Hydroglaciological investigations on the

Lang, H. 1968. Relation between glacier runoff and meteorological factors observed on and outside the glaciers. In: IUGG General Assembly, Berne, *International Association of Scientific Hydrology, Pub.* **No. 79,** 429-439.

Lanzhou Institute of Glaciology and Geocryology. 1980. Study and survey of Batura glacier in the Karakoram mountains. (in Chinese) *Science Press,* Beijing.

Lasaga, A.C. 1984. Chemical kinetics of water-rock interactions. *J. Geophys. Res.,* **89 (No. B6),** 4009-4025.

Lausanne, 1990. International Association of Hydrological sciences, *Pub. No.* **193,** 429-437.

Lemmens, M. 1978. Relation entre concentration en cations dissous et debit de 1'emissaire du glacier de Tsidjiore Nouve (valais), *Catena,* **5,** 227-236.

Lemmns, M.M. and Roger, M. 1978. Influence of ion-exchange on dissolved load of Alpine meltwaters. *Earth Surf. Proce.,* **3,** 179-187.

Lerman, A. 1979. Geochemical processes: Water and sediment environments. Wiley-Interscience. New York.

Liestol, O. 1967. Storbreen glacier in Jotunheimn, *Norway. Nord. Polar. Skr. pp.* 141.

Lliboutry, L. 1983. Modifications to the theory of intraglacial waterways for the case of subglacial ones. *J. Glaciol.,* **29 (No. 102)**, 216-226.

Lorrain, R.D. and Souchez, R.A. 1972. Sorption as factor in the transport of major cations by meltwater from and Alpine glacier. *Quater. Rese.,* **2 (No. 2)**, 253-256.

Luce, R.W., Barlett, R.W. and Parks, G.A. 1972. Dissolution kinetics of magnesium silicates. *Geochim. Cosmochim. Acta,* **36,** 35-50.

Macro Pallis. 1933. Himalayan glaciers. *Himalayan J.,* **6,** 106.

Madden, E. 1847. Notes of an excursion to the Pindari Glacier. *J. Asia. Soc. Beng.,* **17,** 340-450.

Maizels, J.K. 1978. Debit des eaux de fonte, charges sedimentaires et taux d erosion dans le massif du Mont Blanc. *Revue Geograph. Alpine,* **66,** 65-91.

Martinec, J. 1970. Recession coefficient in glacier runoff studies. *International Association of Hydrological Sciences Bull.* **15 (1)**, 87-90.

Mason, K. 1932. The Chong Khumdan Glacier, *Himalayan J.*, **85(No. 1),** 24-35.

Mason, K., Gunn, J.P. and Todd, H.J. 1930. The Shyok flood in 1929. *Himalayan J.*, **2,** 35-47.

Mattes, F.E. 1930. Geologic History of the Yosemite Valley. *U.S. Geol. Surv., Professio. Paper,* **160,** pp. 137.

Mayewski, P.A. and Jeschke, P.A. 1979. Himalayan and Trans-Himalayan glacier fluctuations since AD 1812. *Arct. and Alpine Rese.,* **11(No. 3),** 267-287.

Mayewski, P.A., Pregent, G.P. Jeschke, P.A., and Ahmed, N. 1980. Himalaya and Trans-Himalayan glacier fluctuations and the South Asian Monsoon record. *Arct. and Alpine Rese.,* **12** (No. 2). 171-182.

McCall, J.G. 1960. The flow characterstics of a cirque glacier and their effect on glacial structure and cirque formation, In: Norwegian Cirque Glaciers, Lewis, W.V. (Ed.). *Royal Geogr. Soci. Rese. Series* **4,** 39-62.

Meier, M.F. 1965. Glaciers and climate. In: The Quaternary of the United States. Wright, H.E. and Frey, D.G. (Eds.), Princeton University Press, Princeton, New Jersey, 795-805.

Meier, M.R. 1985. Mass balance of the glaciers and small ice caps of the world. In: *Glaciers, Ice Sheets, and Sea Level: Effects of a* CO_2 *Induced Climate Change (workshop held in Seattle, Washington, September 13-15, 1984,* 139-144.

Menzies, J. 1995. Hydrology of glaciers. In Modern Glacial Environments, Processes, Dynamics and Sediments. Menzies, J. (Ed.), Butterworth-Heinemann, London, pp. 621.

Merrand, Y. and Hallet, B. 1996. Water and sediment discharge from a large surging glacier: bering Glacier, Alaska, USA. *Ann. Glaciol.,* **22,** 233-240.

Metcalf, R.C. 1979. Energy dissipation during subglacial abrasion of Nisqually glacier, Washington, *J. Glaciol.*, **3,** 233-246.

Meybeck, M. 1976. Total mineral dissolved transport by world major rivers. *Hydrol. Sci. Bull.*, **21,** 265-284.

Milliman, J.D. and Meade, R.H. 1983. Worldwide delivery of river sediments to the oceans. *J. Geol.* **91,** 1-9.

Mills, H.M. 1979. Some implications of sediment studies for glacial erosion on Mount Rainier, Washington, *Northwest science*, **53,** 190-199.

Molnar, P. and England, P. 1990. Late Cenozoic uplift of mountain ranges and global climate change: chicken or egg? *Nature,* **346,** 29-34.

Molnar, P. and England, P. 1992. Reply to Summerfield, M.A. and Kirkbride, M.P. (1992) Climate and lanscape response. *Nature,* **355,** 306.

Morel, F.M.M. and Hering, J.G. 1993. Principles and Applications of Aquatic Chemistry. Jon Wiley, New York, pp. 255.

Morland, L.W. and Morris, E.M. 1977. Stress in an elastic bedrock hump due to glacier flow. *J. Glaciol.*, **18,** 67-75.

Moses, C.O., Nordstrom, D.K. Herman, J.S. and Mills, A.L. 1987. Aqueous pyrite oxidation by dissolved oxygen and by ferric ions. *Geochim. Cosmochim. Acta,* **51.** 1561-1571.

Mutreja, K.N. 1986. Applied Hydrology. Tata McGraw-Hill, New Delhi, pp. 243.

Nakawo, M., Fujita, K., Ageta, Y., Shankar, K., Pokhrel, A.P. and Yao, T., 1997. Basic studies for assessing the impacts of the global warming on the Himalayan cryosphere, 1994-1996. *Bull. Glac. Res.,* **14,** 1-6.

Nye, J.F. 1961. The influence of climate variations on glaciers. In: *General Assembly of Helsinki, Snow and Ice Commission, International Association of Hydrological Sciences, Pub.* **No. 54,** 397-404.

Oerlemans, J. 1992. Climate sensitivity of glaciers in southern Norway: application of an energy-balance model to Nigardsbreen, Hellstugubreen and Alfotbreen, *J. Glaciol.*, **38(No., 129)**, 223-233.

Oerter, H. 1984. Der Abfluss an der Pegelstation Vernagtbach (Oetztaler Alpen) 1974-1983. Gesellschaft fur Strahlen-und Umweltforschung mbH, Munich, GSF. Bericht R 363, pp. 27.

Oerter, H., Behrens, H., Hibsch, G., Rauert, W., and Stichler, W., 1980. Combined environmental isotope and electrical conductivity investigations at the runoff of Vernagtferner (Oetztal Alps). Mater. *Glyatsiol. Issled. Khron. Obsuzhdeniya.* **39,**

Ohata, T., Fukushima, Y., Suzuki, M., Motoyama, H., Kawashima, K., and Kubota, H. 1987. Suspended sediment yield in a glaciated watershed of Langtang valley, Nepal Himalaya. *Bull Glacier Res.,* **5,** 19-24.

Ohmura, A. 1981. Climate and energy balance on Arctic Tundra. *Z. Urcher Geogr. Schrift.,* Heft **3**, 448.

Oldham, R. 1904. Note on the glaciation and history of Sind valley, Kashmir. *Rec. Geolog. Surv. of India,* **32,** pt 3, 142-161.

Osterm, G., Bridge, C.W. and Rannie, W.F. 1967. Glacio-hydrology, discharge and sediment transport in the Decade glacier area, Baffin Island, *Geograf. Ann. 49A,* 268-282.

Ostrem, G. 1975. Sediment transport in glacial meltwater streams. In: Glaciofluvial and Glaciolacustrine Sedimentation. Jopling. A.V. and McDonald, B.C. (Eds). Society of Economic Paleontologists and Mineralogists, Pub. **No. 23,** 101-122.

Owens, P. and Slaymaker, O. 1992. Late Holocene sediment yields in small alpine and subalpine drainage basins, British Columbia. In: *Erosion, Debris Flows and Environment. Proceedings of the Chengdu Symposium, July 1992. International Association of Hydrological Sciences, Publ.* **No. 209,** 147-154.

Parks, G.A. 1990. Surface energy and adsorption at mineral/water interfaces: an introduction. *Rev. Mineral.*, **23,** 133-175.

Paterson, W.S.B. 1996. The Physics of Glaciers. Third edition, Pergamon Press, pp. 480.

Phillips, J.D. 1986. Sediment storage, sediment yield and time scales in landscape denudation studies. *Geograph. Analysis,* **18(No. 2),** 161-167.

Puri, 1972. Geological Survey of India, *Ann. Gen. Rep.* **106, No. 2**

Puri, 1974. Geological Survey of India, *Ann. Gen. Rep.* **108, No. 1**

Puri, 1975. Geological Survey of India, *Ann. Gen. Rep.* **109, No. 1**

Puri, 1976. Geological Survey of India, *Ann. Gen. Rep.* **110, No. 1**

Puri, 1977. Geological Survey of India, *Ann. Gen. Rep.* **111, No. 1**

Puri, 1990. Geological Survey of India, *Ann. Gen. Rep.* **124, (No. 8)**.

Radok, V. 1978. Climatic roles of ice. H*ydrol. Sci. Bull.,* **23 (No. 3)**, 333-354.

Rainwater, F.H. and Guy, H.P. 1961. Some observations on the hydrochemistry and sedimentation of the Chamberlin Glacier area, Alaska. *U.S. Geolo. Surv. Prof. Pap.* 414-C.

Raiswell, R., Brimblecombe, P., Dent, D.L. and Liss, P.S. 1980. Environmental Chemistry. Edward Arnold, London.

Raiswell, R. 1984. Chemical models of solute acquisition in glacier meltwaters. *J. Glaciol.,* **30, (No. 104)**, 49-57.

Raiswell, R. and Thomas, A.G. 1984. Solute acquisition in glacial meltwaters. I. Fjallsjokull (south-east Iceland): bulk melt waters with closed-system characteristics. *J. Glaciol.*, **30(No. 104)**, 35-43.

Raiswell, R., Brimblecomble, P., Dent, D and Liss, P.S. 1980. Environmental chemisty. Edward Arnold, London.

Ramaraju, A. 1993. Hydrochemistry of Gangotri glacier, Garhwal Himalaya, India, Ph.D. thesis, Jawaharlal Nehru University, New Delhi, India, Unpublished.

Raymahashay, B.C. 1996. *Geochemistry for Hydrologists*. Allied Publishers Limited. pp. 190.

Raymo, M.E. and Ruddiman, W.F. 1992. Tectonic forcing of Late Cenozoic climate. *Nature.,* **359,** 117-122.

Reynolds, R.C. and Johnson, N.M. 1972. Chemical weathering in the temperate glacial environment of the Northern Cascade Mountains. *Geochimica et Cosmochimica. Acta,* **36(No. 5),** pp. 537-554.

Richards, K.S. 1984. Some observations on suspended sediment dynamics in stoorbregrova, Jotunheim, *Earth Surf. Process. & Landforms,* **9,** 101-112.

Rieger, H.C. 1981. Man versus mountain-the destruction of Himalayan ecosystem. In: *Himalaya. Lall, J.S. (Ed.), Oxford University Press,* 351-376.

Riley, N.W. 1982. Rock wear by sliding Ice. *Ph.D. Thesis,* University of newcastle, U.K.

Robin, G. De. Q. 1976. Is the basal ice of temperate glacier at the pressure melting point?, *J. Glaciol.,* **16**, pp. 183-196.

Rothlisberger, H. 1972. Water pressure in intra-and subglacial channels, *J. Glaciol.,* **11**, 177-203.

Rothlisberger, H. and Lang, H. 1987. Glacial hydrology. In: *Glacio-Fluvial Sediment Transport-An Alpine Perspective.* Gurnell, A.M. and Clark, M.J. (Eds). John Wiley, Chichester, U.K., 207-284.

Sahai, V.K. 1992. Glaciological studies by Geological Survey of India. Proc. Discuss. meet on Ten years Perpective Plan formulation. Pub. WIHG, Deharadun, 9-15.

Sarin, M.M., Krishnaswami, S., Trivedi, J.R. and Sharma, K.K. 1992. Major ion chemistry of the Ganga source waters: Weathering in the high altitude Himalaya, *Proc. Indian Acad. Scie., Earth Planet. Sci.* **101(No. 1)**, 89-98.

Schindler, D.W. 1988. Effects of acid rain on fresh water ecosystems, *Science,* **239,** 149-157.

Schlesinger, M.E. 1986. Carbon dioxide induced changes in seasonal snow cover simulated by the OSU coupled atmosphere-Ocean general circulation model. In: *Snow Watch'85 Report GD-18. Word Data Center A for Glaciology (Snow and Ice),* 249-270.

Sharp, M. 1991. Hydrological influences from meltwater quality data: the unfulfilled potential. *Proc. The BHS 3rd National Hydrological Symposium,* Southampton, 51-58.

Sharp, K.J. 1982a A comparison of landforms and sedimentary sequences produced by surging and non-surging glaciers in Iceland. *Ph.D. Thesis,* University of Aberdeen.

Sharp, M.J. 1982b. Modification of clasts in lodgement tills by glacial erosion. *J. Glaciol.,* **28,** 475-481.

Shoemaker, E.M. 1986. Subglacial hydrology for an icesheet resting on deformable acquifer. *J. Glaciol.,* **32**, 20-30.

Shoemaker, E.M. 1988. On the formulation of basal debris drag for the case of sparse debris. *J. Glaciol.*, **34**, 259-264.

Slatt, R.M. 1972. Geochemistry of meltwater streams from nine Alaskan glaciers. *Goel. Soci. Am. Bull.,* **83 (No. 4)**, 1125-31.

Small, R.J. 1987. Moraine Sediment Budgets. In: *Clacio-fluvial Sediment Transfer-An Alpine Perspective.* Gurnell, A.M. and Clark, M.J. (Eds.), John Wiley, Chichester, UK, 165-197.

Small, R.J., Beecroft, I.R. and Stirling, D.M. 1984. Rates of deposition on lateral moraine embankments, Glacier de Tsidjiore Nouve, Valais, Switzerland. *J. Glaciol.,* **30,** 275-281.

Souchez, R.A. and Lorrain, R.D. 1975. Chemical sorting effect at the base of an Alpine glacier. *J. Glaciol.,* **14, 71,** 261-265.

Souchez, R.A. and Lemmens, M.M. 1987. Solutes. In: *Glacio-fluvial Sediment Transfer-An Alpine Perspective.* John Wiley, Chichester, U.K., 285-303.

Srikantia, S.V. and Padhi, R.N. 1963. Recession of the Bara-Shigri Glacier. In: Symposium on Himalayan Geology. *Geol. Surv. India,* 97-100.

Stern, C.R., Klingfield, K., Schelling, D., Virdi, N.S., Futa, K., Peterman, Z.E. and Amini, H. 1989. The Bhagirathi leucogranite of the high Himalayan (Garhwal India), age petrogenesis and tectonic implications. In: *Tectonics of Western Himalaya.* L. Malincoico jr. and Lillie, R.R. (Eds), Geol. Soci. Am., Special Paper 232.

Stumm, W. and Morgan, J.J. 1981. Aquatic chemistry: an introduction emphasizing chemical equilibria in natural waters. Second edition. John Wiley- Interscience, New York.

Subramaniam, V. 1993. Sediment load of Indian Rivers. *Curr. Sci.,* **64 (No. 11-12)**, 928-930.

Summerfield, M.A. and Kirkbride, M.P. 1992. Climate and landscape response. *Nature* **355,** 306.

Tangborn, W.V., Krimmel, R.M., and Meier, M.F. 1975. A comparison of glacier mass balance by glaciological, hydrological and mapping methods, South Cascade glacier, Washington, In: *Snow and Ice (Proceedings of the Moscow Symposium, August 1971), Johnson, A. Meier, M.F. and Ward, W.H. (Eds), International Association of Scientific Hydrology Pub.* **104,** 185-196.

Tempany, H. and Grist, D.H. 1958. Introduction to Tropical Agriculture. Longman, London.

Tewari, A.P. 1967. Study of the Gangotri Glacier, Uttarkashi District, Central Himalaya, Uttar Pradesh, Rec. of the Geol. Surv. Ind., **106 (No. 2)**, 248-256.

Tewari, A.P. 1971. A short report on glacier studies in the Himalayan mountains by the geological survey of India. Stud. Geomorph. *Carpatho-Balacancia.* **5,** 173-181.

Theakstone, W.H. and Knudsen, N.T. 1981. Dye tracer tests of water movement at the glacier Austre Okstindbreen, Norway. *Norsk Geograf. Tidsskr.,* **35,** 21-28.

Thomas, A.G. and Raiswell, R. 1984. Solute acquisition in glacial melt waters. II. Argentine (French Alps): Bulk meltwaters with open-system characteristics. *J. Glaciol.,* **30,** No. 104, 44-48.

Tranter, M. 1982. Controls on the chemical composition of Alpine glacial meltwaters *Ph.D. Thesis,* University of East Anglia. U.K.

Tranter, M. 1991. Controls on the composition of snowmelt. In: *Seasonal Snowpacks; Processes of Compositional Change.* Davies, T.D., Tranter, M. and Jones, H.G. (Eds), Springer-Verlag, Berlin, 241-27.

Tranter, M., Brown, G., Raiswell, R., Sharp, M. and Gurnell, A. 1993. A conceptual model of solute acquisition by Alpine glacial meltwaters. *J. Glaciol.,* **39. (No. 133)**, 573-580.

Tranter, M. and Raiswell, R. 1991. The composition of the englacial and subglacial component in bulk meltwaters draining the Gornergletscher. *J. Glaciol.,* **37, (No. 125)**, 59-66.

Tranter, M., Raiswell. R and Mills, R.A. 1989. Chemical weathering reactions in Alpine glacial meltwaters. In Mills. D.L., ed. Proceedings of the 6th International Symposium on water-rock interaction, Malvern, 1989, Rolterdam, Balkema, pp. 687-690.

Trimble, S.W. 1975. Denudation Studies: Can we assume stream steady state ? *Science,* **188**: 1207-1208.

Tyndall, J. 1864. On the conformation of Alps. *Philosoph. Maga.,* **28,** 225-271.

Upadhyay, D.S. 1995. Cold Climate Hydrometeorology. New Age International, New Delhi, India. pp. 345.

Vivian, R. 1980. The nature of ice-rock interface: the results of investigation on 20,000 meters square of the rock-bed of temperature glaciers. *J. Glaciol,* **25,** 267-277.

Vohra, C.P. 1971. Gangotri glacier Expedition, Report Geological Survey of India, unpublished.

Vohra, C.P. 1981. Himalayan glaciers. In: The Himalaya. Lall, J.S. and Moddie, A.D. (Eds), Oxford University Press, 138-151.

Vohra, C.P. 1989. Gangotri Glacier, *Indian Mountaineer.* Published by the Mountaineering Foundation of India, New Delhi.

Vohra, C.P. 1993. Himalayan glaciers. In: *Harnessing the eastern Himalayan river.* Ramaswamy Iyer (Ed.), 56-66. Konark publishers Pvt Ltd. New-Delhi.

Walder, J.S. 1986. Hydraulics of subglacial cavities. *J. Glaciol.,* **32(No. 112)**, 439-445.

Walling, D.E. 1983. Sediment delivery Problem. *J. Hydro.,* **65,** 209-237.

Walling, D.E. and Kleo, A.H.A. 1979. Sediment yield of rivers in areas of low precipitation: a global view. In: *The Hydrology of Areas of Low Precipitation. International Association of Hydrological Sciences, Pub.* **No. 128**, 479-493.

Walling, D.E. and Webb, B.W. 1988. The reliability of rating curve estimates of suspended sediment yield: some further comments. *International Association of Hydrological Sciences,* Pub. **No. 174**, 337-350.

Walker, H., and Pascoe, E.H. 1907. Notes on certain glaciers in Lahul. *Rec. Geol. Surv. India,* **35pt** 4, 139-147.

Warburton, J. 1990a. Comparison of bed load yield estimates for a glacial meltwater stream. In: Hydrology in Mountainous Regions. I-Hydrological measurements: The Water Cycle. Proc. Two Lusanne Symposia, August 1990. International Association of Hydrological Sciences, Pub. **No. 193**, 315-322.

Warburton, J. 1990b. An Alpine proglacial fluvial sediment budget. Geograf. Ann., **72A(No. 3-4)**, 261-272.

Watanabe, O., Iwata, S. and Fushimi, H. 1986. Topographic characteristics in the ablation area of the Khumbu glacier, Nepal Himalaya. *Ann. Glaciol.,* 7, 177-180.

Williams, R.B.G. 1986. Intermediate statistics for geographers and earth scientists. Macmillan education, London, pp. 712.

Wollast, R. 1967. Kinetics of alteration of K-feldspar in buffered solutions at low temperature. *Geochim. Cosmochim. Acta.,* **31,** 635-648.

Yamada, T. 1998. Glacier lake and its outburst flood in the Nepal Himalaya. Monograph no. 1. Data center for Glacier Reserch. *Japanese Soci. Snow and Ice* pp. 96.

Yamada, T. 1992. Report for the first research expedition to Imja glacier lake WECS Report no. 3/4 120892/1/1, Seq. No. 412.

Yamada, T. 1993. Glacier lakes and their outburst floods in the Nepal Himalaya. Pub. *Water and Energy Commission Secretariat, Kathmandu, Nepal,* pp. 37.

Yamada, T. 1998. Glacier lakes and its outburst flood in the Nepal Himalaya. Monograph No. 1. Data Center Glac. Rese. *Jap. Soc. Snow & Ice,* pp. 96, Japan.

Yao Tandong and Shi Yafeng 1990. Fluctuations and future trend of climate, glaciers and discharge of Urumqi River in Xinjiang, *Science in China,* **33 (No. 1),** *series B.*

Young, G.J. 1982. Hydrological relationship in a glacierised mountain basin. In: *Hydrological Aspects of Alpine and High Mountain Areas, Proc. Exeter Symp., Glen, J.W. (Ed.), International Association of Hydrological Sciences, Pub. 128.*

Young, G.J. 1985. Techniques for the Prediction of Runoff from Glacierized Areas. *International Association of Hydrological Sciences,* **No. 149**, 149.

Young, G.J. 1990. Glacier Hydrology. In: Northern Hydrology Symposium, July 1990, Saskatoon National Hydrology Research Institute Symposium, Prowse, T.D. and Ommanney, C.S.L. (Eds), **No. 6**, 35-162.

Young, G.J. and Hewitt, H. 1990. Hydrology Research in the Upper Indus Basin, Karakoram Himalaya, Pakistan. Hydrology of Mountainous Areas (Proc. Strbske Pleso workshop Czechoslovakia). International Association of Hydrological Sciences, Pub. **No. 190**, 139-152.

Zheng, B. 1989. The influence of Himalayan uplift on the development of Quaternary glacier. *Z. Geomorphol.* Suppl. Bd., **76,** 1-89, 51-59.

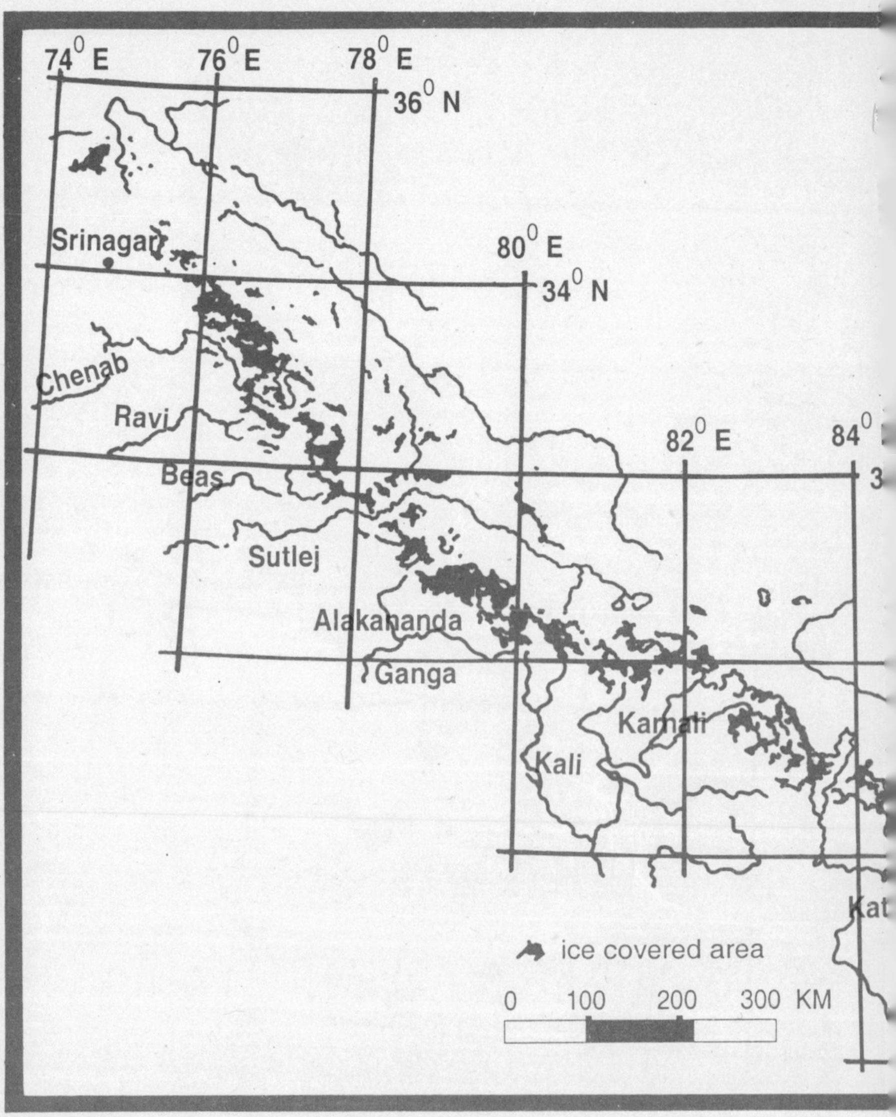

74° E
76° E
78° E
36° N
80° E
34° N
82° E
84°
Srinagar
Chenab
Ravi
Beas
Sutlej
Alakananda
Ganga
Karnali
Kali
Kat
ice covered area
0 100 200 300 KM